STATE OF IDAHO
ARNOLD WILLIAMS, Governor

IDAHO BUREAU OF MINES AND GEOLOGY
A. W. FAHRENWALD, Director

GOLD in IDAHO

By
W. W. STALEY

University of Idaho
Moscow, Idaho

TABLE OF CONTENTS

ILLUSTRATIONS

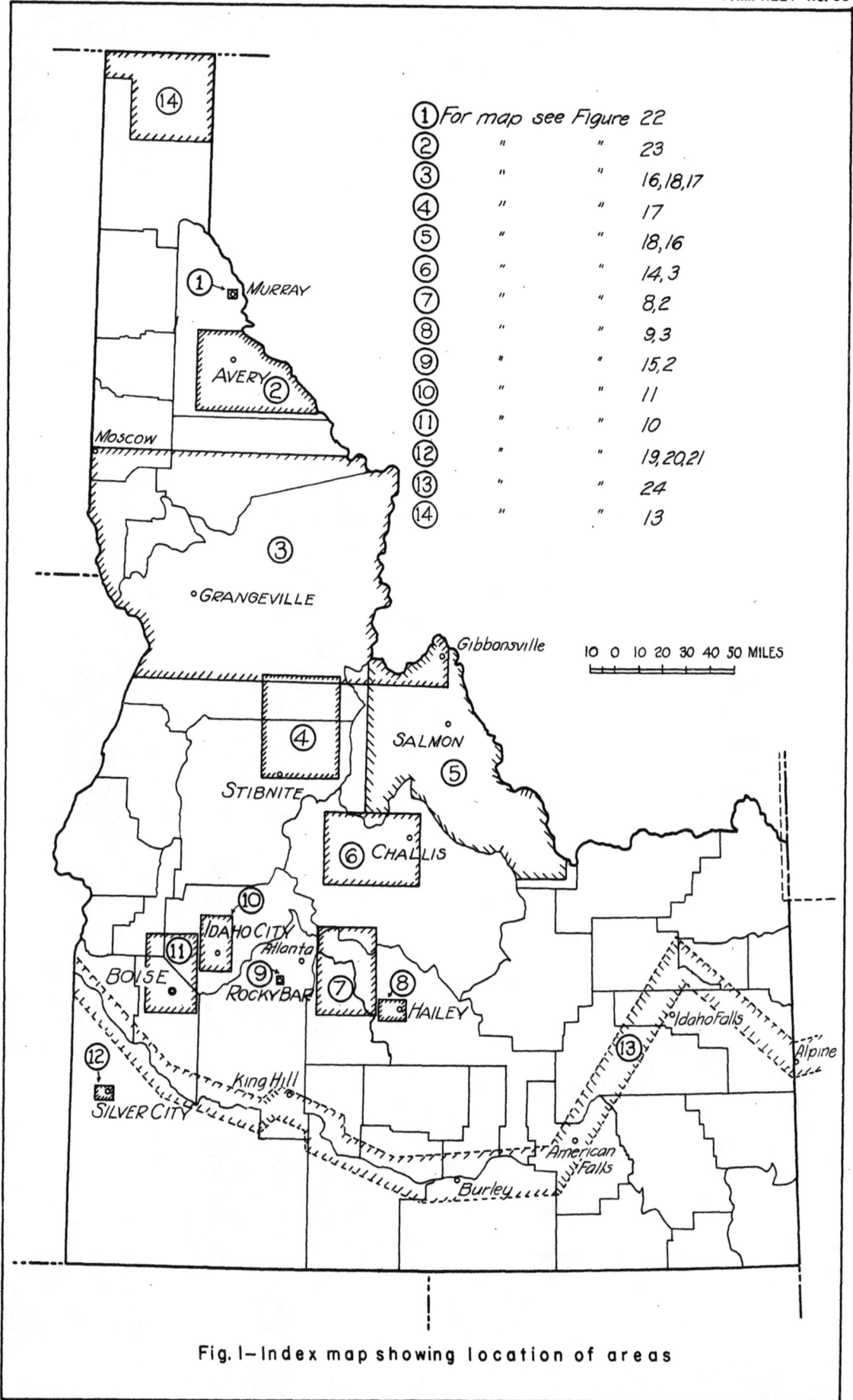

① For map see Figure 22
②   "   " 23
③   "   " 16,18,17
④   "   " 17
⑤   "   " 18,16
⑥   "   " 14,3
⑦   "   " 8,2
⑧   "   " 9,3
⑨   "   " 15,2
⑩   "   " 11
⑪   "   " 10
⑫   "   " 19,20,21
⑬   "   " 24
⑭   "   " 13

Fig. I—Index map showing location of areas

GOLD *in* IDAHO

By W. W. STALEY

▼

INTRODUCTION

The purpose of this paper is the presentation of data and information which will promote the production of gold in the state when gold mining is again permitted to flourish.

Since the discovery of gold at Pierce City in 1860 there have been a great many mining operations carried on. Most of them have risen to their peak and then died out. It is the writer's belief that the lack of ore was not always responsible. Many conditions influenced early day mining—poor roads, if there were any at all; hostile Indians; the desire for a quick stake; development of our extensive irrigating systems; very inefficient metallurgical processes (if the ore was not free milling, very little could be done with it in the early days and many old tailings dumps might well be investigated); and a swarm of unscrupulous promoters descending upon an infant industry may be mentioned. Consequently, many discoveries were not brought to the prospect stage; and good prospects were left undeveloped. The Director of the mint report for 1883 lists, in Idaho, 240 known operations. Today most of these particular ones are unknown. The reports on these early-day discoveries were no less enthusiastic than those we read today. Values given in the various mint reports run from a few dollars to several thousand dollars per ton.

There is a fairly extensive literature on the history and geology of the leading districts in these early mint reports. At a later date Waldemar Lindgren contributed greatly to our geological and mining knowledge. Since about 1915 many geological papers have appeared. It is with the reports previous to 1906 that we are mostly concerned. The reason is to bring to the prospector's attention the fact that there are, apparently, many areas favorable to investigation that have been forgotten for the past 40-50 years. Some of the areas containing the rich production of years ago (for example, the Silver City district in Owyhee County) could well stand further examination.

A considerable number of maps are reproduced in this report. In many cases later geological maps may be found. Invariably, however, the more recent work includes insufficient data on the early operations. As the main purpose here is to bring to light the old timer's efforts, emphasis is given to the older reports, even at the expense, in many instances, of more scientific data. So far as the geological evidence is concerned there is, from a practical standpoint, little essential difference between Lindgren's reports and those 30 years later.

In assembling the production figures difficulties were encountered regarding the accuracy of the period 1860 to 1880. For example, in 1866 three sources gave 401,200, 500,000, and 850,000 ounces. The largest figure is Wells, Fargo and Co., and the smallest that of the mint. Since 1880 the reports have been by counties and are apparently accurate. Placer data preceding 1905 is practically nonexistent.

HISTORY

Gold is supposed to have first been found in Idaho in 1860 by a Capt. Pierce at what is now Pierce, in Clearwater County (1). At the time of discovery Pierce City was supposed to be in Shoshone County. More likely it was in Nez Pierce County. (See Plate 1.) The Oro Fino and Elk City discoveries were made about the same time (1). In 1861 the Salmon River placers were found (1). These deposits begin at about Whitebird and continue on up the river to about Shoup. (See Fig. 16.) Florence, Warren, and Buffalo Hump were found in 1862. A party led by George Grimes discovered the Boise Basin in August of 1862 at the spot where Centerville now stands (2). It would appear that the credit should go to Moses Splawn (3).

These discoveries were made before Idaho was officially created as a territory (4). The territory at that time included the area shown for 1864 in Plate 1. In the spring of 1864 she lost the greater part of her area through the formation of Montana. Later, Idaho lost the greater portion of Oneida County, as shown in 1864, to Wyoming. The eastern boundary thenceforth remained about as it is at the present time. By the summer of 1863 there were supposed to be 15,000 to 20,000 men in the Boise Basin.

The Snake River placers were first prospected as early as 1862 at Jackson Hole on the headwaters of the Snake (5) which was at that time in what would later become Idaho Territory. It was not until the late sixties and early seventies that serious mining began along the Snake River.

On May 18, 1863 the Jordan Creek placers on Jordan Creek, near Booneville, Owyhee County, were found (6).

Placer gold was found on the Boise River at Rocky Bar in the same year.

_____
1. See references at the end of pamphlet.

The first difficulty one is confronted with when trying to locate early-day mining activities is the rapid change in county names and boundaries. Because of this, wrong impressions are gained as to the production of certain counties. For example, the production for Nez Perce County would be taken at about 37,000 oz. On investigation we find that after the formation of Clearwater County the production of Nez Perce dropped to a hundred or so ounces from 1911 on, while Clearwater produced at about the former rate of Nez Perce. The answer is, of course, that the producing area is not in what is at present Nez Perce County. Some of the Shoshone County total came from what is now Clearwater County. This same confusion arises with Shoshone, Idaho, Boise, Alturas (now made up of Blaine, Custer, Elmore, Fremont, Butte, Lincoln, part of Lemhi, etc., Counties), Owyhee, and Oneida. Ada has also undergone many changes in boundary. This is not the place for a history of the counties in Idaho, but some guidance is necessary to gain an understanding of the location of mining areas.

Plate 1 shows an approximate development of the counties. When the territory of Idaho was created, the Washington Territorial Legislature had already formed and named four counties in what was to later become Idaho. They were: Shoshone, Nez Perce, Idaho, and Boise Counties (7).

The session of the first Idaho Legislature was at Lewiston from December 7, 1863 to February 4, 1864. During this meeting three new counties were formed. Namely, Owyhee, Alturas, and Oneida. (On Plate 1, map 1865 really shows Idaho as the counties existed in 1863. The map 1864 existed only for a few months until Congress formed Montana. The Idaho Legislature, however, had already bounded and named the western Montana counties (8).)

Plate 1* gives a general idea of the growth from four named counties and one nameless one (the present panhandle) to the present total of 44. The dates identifying the sketches do not necessarily represent the exact dates of formation of new counties. They are probably the dates of the publication in which the originals appeared. There is very much of an overlap in most cases. However, early descriptions may with fair accuracy be placed in present counties.

Ada County was created at the second session, December 22, 1864.

Lemhi County was created January 9, 1869.

Up to and including 1870 there were nine counties in the territory. From then until about 1919 there was a constant formation of counties with frequent changing of names and boundaries.

Quartz mining usually began a year or so after the discovery of the placers. It ceased if the terri-

tory failed to build roads and, also, when the ore became too refractory for amalgamation, which was usually at relatively shallow depths. A great many of the roads were toll roads.

To give some insight into early-day mining, a synopsis of the Director of the Mint report for 1882 is reproduced (9). The only reason for choosing this particular year was that it is more voluminous than other years. His report on Idaho for 1882 consisted of the equivalent of 28—8½ by 11 typewritten pages, single spaced. In addition, he listed some 240 mining operations by organization name. Many of these were lead-silver mines, especially those around Hailey, Bay Horse, and southern Lemhi County. The counties contained these companies as follows: Alturas, 120; Lemhi, 5; Custer, 57; Owyhee, 35; Boise, 15; Washington, 4; and Idaho, 3. Alturas County at this time contained the Rocky Bar and Atlanta gold areas. With the exception of eastern Custer County and the Wood River region at Hailey, practically all operations were gold mines. The silver production for this year exceeded the gold by $500,000.

ALTURAS COUNTY

(Western Alturas County is now Elmore County.)

At Rocky Bar are listed the Idaho and Vishnu, and the Ada Elmore. (See Fig. 2.)

In the Atlanta district are mentioned the Buffalo, Monarch, Last Chance, and Leonora. (See Fig. 2.)

Very optimistic statements are made concerning the grade of the ore in these districts. Values between $60 and $2000 per ton are offhandedly spoken of.

The greater part of the descriptive material is devoted to the eastern part of Alturas County around Hailey. (See Fig. 3.) This was a very productive lead-silver area. North of Galena, in what was called the Sawtooth district (Fig. 2), high-grade silver mines are described. One running several hundred ounces per ton is enthusiastically reviewed. Some placer operations were active here.

Little Wood River district is described in detail. It was noted for lead-silver production. There were several smelters, mills, and a very considerable mining operation. Again the ore is said to run a hundred or so ounces of silver per ton.

The production of Alturas County for 1882 is given at $175,000 gold and $770,000 silver.

BOISE COUNTY

It is stated that a belt of rich mines, commencing at Gold Hill, extends northeast for a distance of about 40 miles.

The Banner district, in the northeastern part of the county, contained quite a number of oper-

---

* The originals were obtained through the courtesy of the University of Idaho Library.

(F.J.Scott: Mining Regions of Idaho, 1882)

Scale in Miles
5 4 3 2 1 0   5   10   15

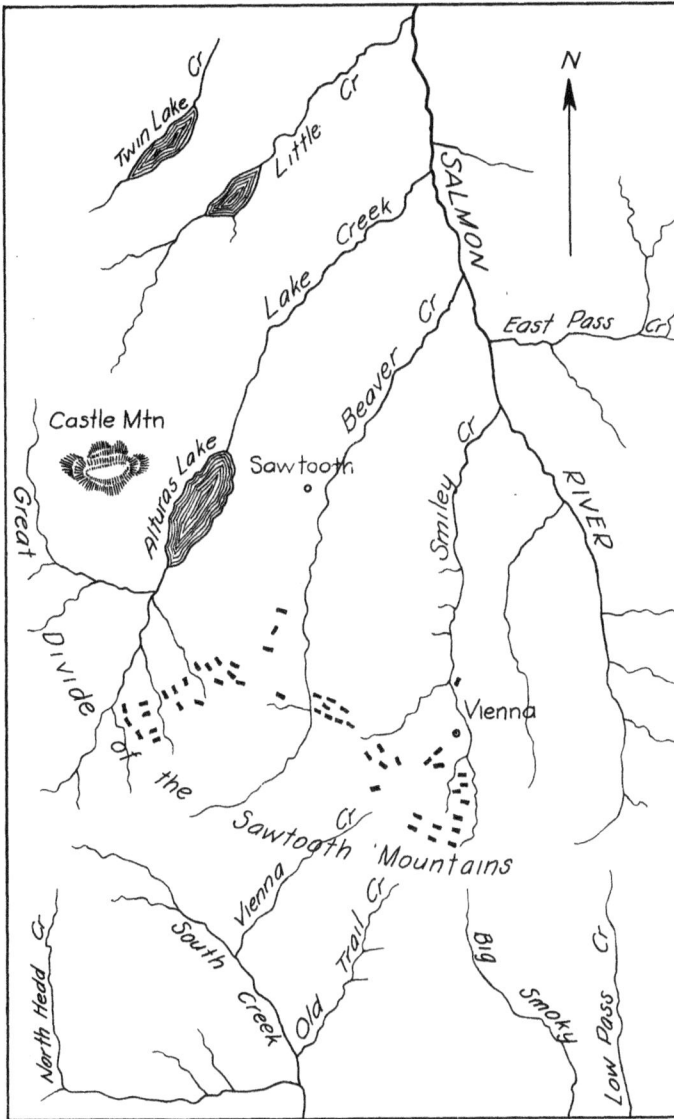

The Sawtooth Mining District

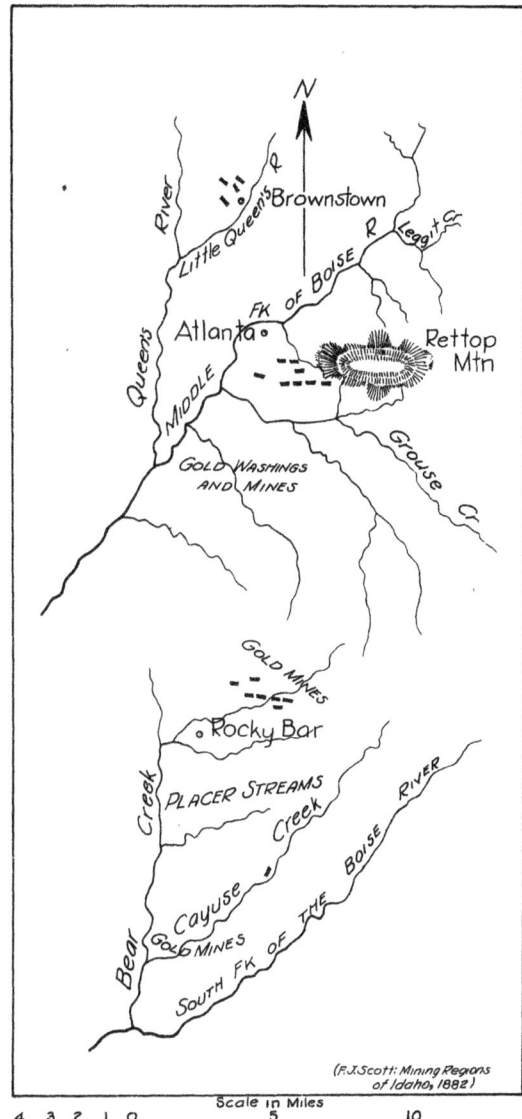

(F.J.Scott: Mining Regions of Idaho, 1882)

Scale in Miles
5 4 3 2 1 0   5   10   15

Atlanta and Rocky Bar

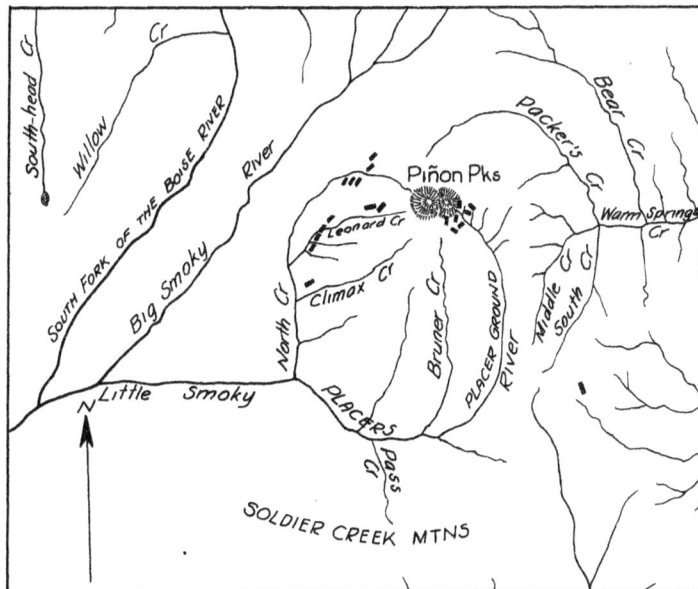

(F.J.Scott: Mining Regions of Idaho, 1882)

Smoky District
Figure 2.

ating properties. Among these are the Banner mine; the Crown Point; and the Deer Lodge (reporting assays of over $300 per ton). The Monarch, Homestake, and Daisy report assays of the same order. The Gold Hill Mining and Milling Co. had been developed to a depth of 400 ft. Many other placer and lode mines are described.

Production for 1882 was $290,000 gold and $20,000 silver.

CUSTER COUNTY

The Bay Horse district and most of eastern Custer County contains lead-silver mines. (See Fig. 3.)

In the Yankee Fork district are found the gold producers. They are on Yankee Fork, Jordan Creek, Estes Peak, Norton Hill, and Custer Hill. (See Fig. 3.) The Custer mine is said to have produced, up to this time, $2,000,000. In 1881 the production was $850,000.

A considerable number of properties are discussed, both placer and lode.

During 1882 Custer County produced $300,000 gold and $950,000 silver.

IDAHO COUNTY

Very few new discoveries since the previous report were noted. Placer mining was dropping off, and, to a large extent, was conducted by the Chinese. It is suggested that there are a number of likely prospects.

Most of the quartz mining was in the Warren district.

It is interesting to note the following items:

"Placer mining in Warren is mostly done by Chinamen. Five companies own over a mile of the creek bottom, and employ about 200 men. They have reported the following production:

"Took Sing Company	$14,120
Lin Wo Company	21,500
Hung Wo Company	17,400
Wing Wo Company	15,000
Shun Lee Company	11,260
Total	$79,280

"Other small Chinese companies produced $22,500, white men $12,500, and individual Chinamen about $1,000, making the total production of the district $126,450."

The aggregate production of Idaho County for 1882 is given as follows:

Florence district	$ 35,000
Salmon River mines	50,000
Warren district placer	115,280
Quartz	11,170
Elk City and scattering	35,000
Total	$246,450

Of this $240,000 was estimated as being gold.

LEMHI COUNTY

"Lemhi County, lying along the eastern border of northern Idaho, is fast becoming known as a region rich in mineral wealth." So writes the Director of the Mint for 1882. He points out that much of the county's most promising area was transferred to Custer County. For example, in 1881 the Mount Estes mines were in Lemhi County.

Leesburg district.—Located about 20 miles west of Salmon City. Placer gold was discovered in 1866. The deposits were very rich but limited in extent. Several lode mines having $20-$40 ore are mentioned. Transportation difficulties were encountered.

Gibbonsville and Mineral Hill (below Salmon City on the Salmon River) are the other gold districts mentioned. In the Spring Mountain and Texas districts lead-silver mines were operating.

The production for 1882 was $180,000 gold and $30,000 silver.

OWYHEE COUNTY

Mining activity in this county (Jordan Creek placers) is among the earliest in the state. The production is essentially gold-silver, the values of each being about equal. Many famous mines had operated and were slowly returning to peak production. The War Eagle Mountain, Florida Mountain, Flint, DeLamar, and Mammoth districts all contained producing properties. Many of them, with their output, are discussed at length. Ore ran from a substantial amount up to as much as $200 per ton.*

The production for 1882 was $200,000 gold and $230,000 silver.

WASHINGTON COUNTY

The activity was almost entirely prospecting. Copper ore carrying gold and silver values was responsible for most of the interest. Transportation difficulties were involved and for that matter are still a handicap.

The foregoing outline is but a brief repetition from the original article. It gives some idea, however, of the mining interest some 60 years ago. Many of the properties mentioned have been forgotten. Under present conditions they might well become profitable producers.

---

* It is interesting to read Bancroft's History of Washington, Idaho and Montana on the values of these ores.

PRODUCTION

The total production of gold before 1934 has been estimated as high as $300,000,000. This would be approximately 15,000,000 ounces. Records and estimates by competent authorities show but a little more than half this from 1863 to 1942. The doubtful period is the time previous to 1870, especially before 1866 and 1863. A careful search of the records gives the data which will appear from time to time in this report. For the years 1866 to 1869, inclusive, there is a difference of estimate amounting to 873,000 oz. There must have been a considerable amount between 1860 and 1863, although there are no factual records (10). It is informative to quote the following from Bancroft:

"Something should be said of the precious metals, whose existence in Idaho caused its settlement. The standard of gold bars being 1,000, anything below half of that was denominated silver. A bar 495 fine was 500 fine of silver, worth $10.23¼ per ounce; a bar 950 fine was 45 fine of silver, and was stamped $19.63 per ounce, as in the case of the Kootenai gold. Santiam gold (Oregon) was 679 fine; Oro Fino gold-dust assayed $16 to the ounce; Elk City from $15.75 to $16.45; Warren's Diggings $10.08 to $14.54; Florence from $11.80 to $13.75; Big Hole (Montana) $17.30; Beaver Head $18.37 to $18.50; Boise $14.28 to $17.40, little of it assaying less than $15, at which price the merchants of Idaho City pledged themselves to take it, while paying only $10 for Owyhee and $12 for Florence. (Boise News, Nov. 3, 1863 and Jan. 23, 1864.) The actual amount of gold produced in any particular district of either of the territories for a given time would be difficult of computation, and only approximate estimates can be made of the amount carried out of the country by individuals or used as a circulating medium in trade, and gradually finding its way to the mints of Philadelphia or San Francisco. Without vouching for the correctness of the estimation, I shall quote some from the discovery of the Clearwater mines for several years thereafter. The Portland Oregonian of Jan. 18, 1862, gives the amount brought to that city during the previous summer and autumn as $3,000,000, but this was not all Idaho gold, some being from Oregon mines. G. Hays, in Ind. Aff. Report for Oct. 1862, says, 'I should think between $7,000,000 and $10,000,000 a fair estimate' for the gold taken from the Nez Perce mines in two years. In six months from June to November 1863, the express company shipped $2,095,000, which was certainly not more than a third of the product of the Idaho mines alone. The Idaho World of June 30, 1866, placed the product of Idaho and Montana for 1865 and 1866 at $1,500,000 monthly. See also U. S. Land Off. Rept., 1865, 15, corroborating it. J. Ross Browne, in his Mineral Resources, gives the following figures for 1866: Montana $12,000,000, Idaho $6,000,000, Oregon $2,000,000, and Washington $1,000,000; but the S. F. Chronicle makes the product of Idaho for 1866 $8,000,000; for 1867 $6,500,000; for 1868 $7,000,000; for 1869 $7,000,000; for 1870 $6,000,000; for 1871 $5,000,000, suddenly dropping in 1872 to $2,514,090. None of these figures can be depended upon, the government reports least of all; but they enable us to make sure that Idaho and the twin territory of Montana had furnished the world a large amount of bullion without yet having begun in earnest to develop their mineral riches."

Bancroft (11) further says:

. . . But the report of the mint director is no more than a guide to the actual amount of gold produced, the larger part of which is shipped out of the territory by banking firms or in private hands, and goes to the mint at last without any sign of its nativity. The total gold product of Idaho down to 1880 as deposited at the mints and assay offices has been set down at $24,157,447 and of silver $727,282.60. But some $60,000,000 should be added to that amount, making the yield of precious metals for Idaho $90,000,000 previous to 1881, when the revival of mining took place. . . . The Virginia and Helena Post of Jan. 15, 1867, makes the output of the Idaho mines in 1866 $11,000,000. When J. Ross Browne made his report to the government on the yield of the Pacific states and territories he omitted Idaho, which had produced from $10,000,000 to $20,000,000 annually for 4 years. Silver City Avalanche, Feb. 9, 1867 . . .

The statistics on gold production for this paper were obtained from the following sources:

1863	Estimate by Waldemar Lindgren, 18th Annual Report, U. S. Geol. Sur., vol. 3, 1897, p. 652.
1864 to 1868	Reports on Mineral Resources, by J. Ross Browne, House Executive Documents.
1869 to 1875	Mineral Resources West of the Rocky Mountains, by R. W. Raymond, House Executive Documents.
1876 to 1879	No yearly report. Estimates obtained from mint reports.
1880 to 1904	Production of Precious Metals, Dept. Documents, Director of the Mint.
1905 to 1942	U. S. Geol. Sur. Mineral Resources to 1925, and after this date the U. S. Bur. of Mines Mineral Resources.

Wood River District map:

Boyles Mtn
Dane Cr
Lake Cr
Trail Cr
Smelters
Ketchum
Warm Spr
Lewis Cr
North Star Cr
Greenhorn Cr
East Fork
Deer Cr
Indian Spr Cr
Bullion Mtn
Horse Cr
Bullion
Hailey
Kelley Cr
Elk Cr
Gilman
Croy
W. Rock Cr
Bellevue Cr
Broadford
Bellevue
Buck Cr
Sinks
Camp Creek
Poison Cr
Rock Cr
WOOD RIVER
Silver Creek

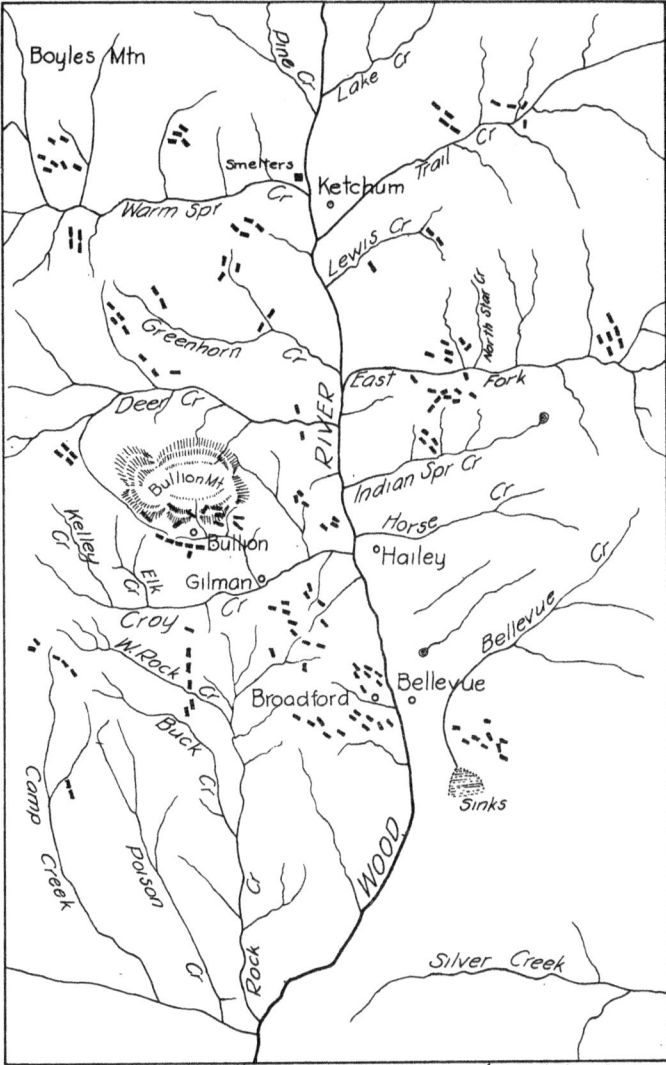

Wood River District   (F.J. Scott: Mining Regions of Idaho, 1882)

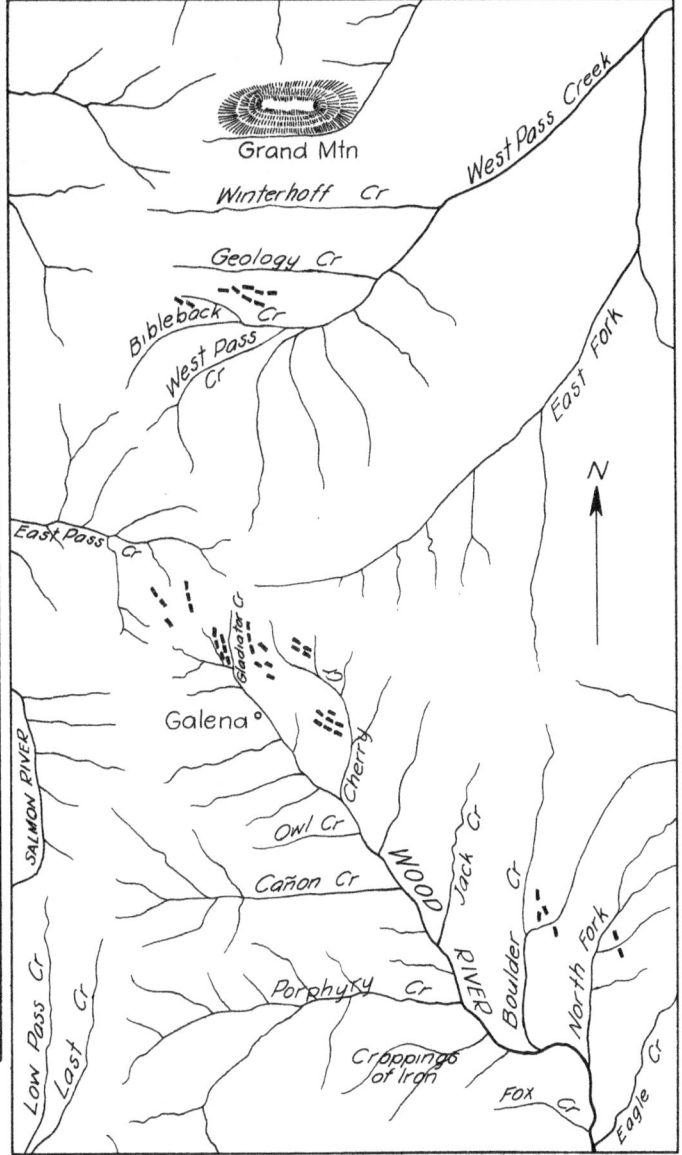

Upper Wood River map:

Grand Mtn
West Pass Creek
Winterhoff Cr
Geology Cr
Bibleback Cr
West Pass Cr
East Fork
East Pass Cr
Gladiator Cr
Galena
Cherry
Owl Cr
Jack Cr
Cañon Cr
Boulder Cr
Porphyry Cr
WOOD RIVER
North Fork
SALMON RIVER
Low Pass Cr
Last Cr
Croppings of Iron
Fox Cr
Eagle Cr
N

Upper Wood River   (F.J. Scott: Mining Regions of Idaho, 1882)

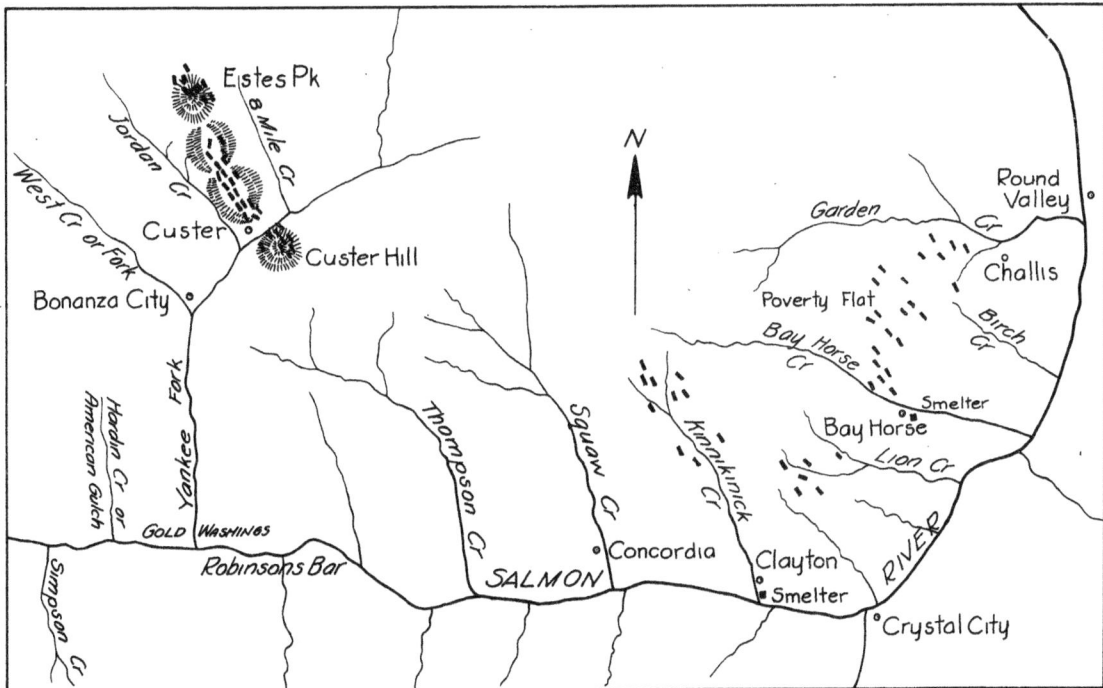

Yankee Fork and Bay Horse Districts map:

Estes Pk
8 Mile Cr
Jordan Cr
West Cr or Fork
Custer
Custer Hill
Bonanza City
N
Round Valley
Garden Cr
Poverty Flat
Challis
Birch Cr
Bay Horse Cr
Hardin Cr or American Gulch
Yankee Fork
Thompson Cr
Squaw Cr
Kinaiknick Cr
Smelter
Bay Horse
Lion Cr
Gold Washings
Robinsons Bar
SALMON RIVER
Concordia
Clayton
Smelter
Crystal City
Simpson Cr

Yankee Fork and Bay Horse Districts   (F.J. Scott: Mining Regions of Idaho, 1882)

Figure 3.

It is very seldom that the various overlapping sources of this data checks each other. For example, the mint reports seem usually to be about 11 per cent lower than other government information. For this reason the years 1876-77-78 have been increased. The mint report was about 11 per cent low compared to the precious metal reports for years on each side of the questionable period. The year 1879 is missing for some unknown reason; an arbitrary value of the average of the two adjoining years was assumed.

The sources cited above recommended for the years 1866-67-68-69, 401,200, 325,000, 350,000, and 350,000 oz. respectively. On the other hand, they stated that Wells, Fargo and Co., whom they admitted were well qualified to make an opinion, as they were the shipping agents, reported the following for these years: 850,000, 450,000, 500,000, and 500,000 oz., respectively. One other source estimates 500,000 oz. for 1866. The totals reproduced herein are based on the lower set of figures, although it would seem as reasonable to use Wells,

Fargo and Co., for, as R. W. Raymond said in his report for 1868, they should know just what they had handled. Nevertheless he very substantially reduced their figures.

Individual county data, with the exception of Boise County, are not available before 1880.

Plate 2 includes a graph showing the total production and the placer production from 1900 to 1942.

The distinction between lode and placer output would have seemed to be made consistently after 1904, especially so far as counties are concerned.

The following table gives the total production of gold as obtained from the sources listed above. Later, under each county the county production will be listed. It has already been shown that Bancroft suggests the Idaho total output, as given by the government records, to be substantially low. There is no way to correct the situation so only the actual recorded figures are included in the table.

TOTAL PRODUCTION OF GOLD IN IDAHO, OUNCES

1863—350,000	1879— 81,400[6]	1895—125,517	1911— 66,389	1927— 15,315
1864—323,000	1880— 99,435	1896—112,409	1912— 66,816	1928— 20,981
1865—329,100	1881— 85,000	1897—102,813	1913— 65,043	1929— 20,247
1866—401,200[1]	1882— 75,000	1898— 91,689	1914— 55,744	1930— 21,446
1867—325,000[2]	1883— 70,000	1899—102,106	1915— 57,069	1931— 18,362
1868—350,000[3]	1884— 62,500	1900—100,525	1916— 53,977	1932— 46,886
1869—350,000[4]	1884— 93,742	1901— 92,750	1917— 38,933	1933— 64,592
1870—300,000	1886— 86,980	1902— 73,047	1918— 33,999	1934— 84,817
1871—250,000	1887—104,669	1903— 83,737	1919— 34,503	1935— 83,821
1872—134,800	1888— 98,003	1904— 82,739	1920— 23,419	1936— 80,291
1873—125,000	1889— 99,445	1905— 52,033	1921— 26,400	1937— 81,861
1874—100,000	1890— 82,080	1906— 55,588	1922— 24,256	1938—103,513
1875— 64,000	1891— 81,540	1907— 60,755	1923— 37,124	1939—116,900
1876— 41,000[5]	1892— 83,271	1908— 69,827	1924— 26,922	1940—146,480
1877— 82,500[5]	1893— 81,930	1909— 70,329	1925— 20,887	1941—149,816
1878— 63,300[5]	1894—111,687	1910— 53,060	1926— 13,669	1942— 95,020

[1] Other sources give 500,000 and 850,000 for 1866.
[2] Other sources give 450,000.
[3] Other sources give 500,000.
[4] Other sources give 500,000.
[5] Mint report increased 10% to give the figure used.
[6] Average of 1878 and 1880. No other data available.

The total of the above tabulation is 8,110,004 oz. If the maximum value is used, the total becomes 8,883,840 oz. (See page 8 for comments by Bancroft.)

PLACER GOLD PRODUCTION IN IDAHO, OUNCES

1900—43,721	1909—13,661	1918—13,371	1927— 7,520	1936—34,430
1901—36,461	1910—11,782	1919— 9,228	1928— 8,192	1937—40,540
1902—34,547	1911—19,599	1920— 5,506	1929— 4,130	1938—54,079
1903—36,231	1912—30,524	1921— 8,755	1930— 3,988	1939—48,663
1904—23,849	1913—33,575	1922— 8,900	1931— 5,214	1940—60,409
1905—16,470	1914—33,885	1923—24,125	1932—12,440	1941—72,395
1906—17,100	1915—28,294	1924—17,324	1933—23,290	1942—44,580
1907—17,270	1916—21,725	1925—12,693	1924—27,256	
1908—13,822	1917— 6,542	1926— 8,361	1935—31,751	

The total production of placer gold since 1900 is 1,026,198 oz. For the same interval, 1900 to 1942, the total production was 2,575,888 oz. Thus, about 40 per cent was placer.

The General Geology of Gold Deposits

According to Emmons (12), Bateman (13), and Newhouse (14) (many other authorities and reports might be mentioned) practically all of the gold deposits of the world are associated with acidic igneous rocks; there are a few exceptions, most of which are of minor importance. (An inspection of the opinions expressed in the literature for Idaho shows that her deposits are no exception to this theory).

By acidic igneous rocks are meant the diorites, monzonites, rhyolites, andesites, trachytes, granodiorites, and granites (15). Granodiorite or granite is the usual rock. (In Idaho, the Idaho batholith, see Plate 2, occurs in various phases of granite and granodiorite.) With very few exceptions the ore deposition takes place within a mile of the contact between the granite and the invaded rocks. Where the country rock is limestone very few gold deposits of importance have occurred. Islands of country rock in the invading granite are favorable places to search for deposits, as are "islands of granite" in the country rock.

The country rock, that is, the rock into which the above named igneous rocks are intruded or in contact with, may consist of: limestone, shale, slate, conglomerate, andesite, rhyolite, granite, basalt, quartzite, schist, gneiss, etc. (16).

The mechanics of the intruding magma, its cooling and solidifying, and the escape of gases and vapors causes distortion of the surrounding rocks and distortion of the solidfying igneous rock at the contact. This distortion takes the form of anticlines, synclines, folds, monoclines, rolls, and domes (17). Cooling and the escaping gases cause these structures to become breciated, jointed, fractured, fissured, and faulted with also the formation of shear zones. In the open spaces thus made the mineralizing solutions circulate and deposit their mineral content. The deposition may be at the crest, on the flanks, or the bottom trough of an inclined structure, and it may depart from the main channel and follow cross openings (joints, shear zones, etc.). Some rocks, because of their physical characteristics, give fine, clean cut fractures; others splinter and provide a series of more or less parallel fractures or slips. As a rule, the latter is more favorable for commercial bodies of ore. The more surface presented to the ascending solutions the better the chance that their mineral load will be deposited. A preponderance of the gold deposits are of the replacement type in open fissures (18).

Emmons states that the association of auriferous lodes with igneous rocks is practically universal, and further points out that they are chiefly associated with the acidic rocks (as previously named) and occur in certain definite positions with relation to the intruding magma (19).

If erosion or glaciation has been intense or prolonged an originally extensive vein system may be eroded to its "roots", thus leaving very little or no commercial ore. (See Fig. 4.)

Figures 5 (20) and 6 (20) show the relationship between the intrusive and the country rock.

It will be noted from Fig. 5 that the deposits are located in the **roof** and the **hood.** Practically no valuable deposits are ever found **below** the **dead line.** The hood is that part of the batholith between the dead line and the contact with the intruded rock. As the depth below the original surface increases, the dead line converges closer to the contact.

In Fig. 6 the deposits are concentrated in and around the cupolas or bulges. While gold is found at the summit and intermediate cupolas, the greatest concentration is at the trough cupolas (20).

Plate 2 shows a close approximation of the outline of the Idaho batholith (21). On this plate all granitic and related rocks (granodiorite and monzonite), and some small areas of granites of possibly different age (later age?) have been lumped together and called the Idaho batholith. The invaded rocks consist of local gravel areas, extensive slates, quartzites, schists, gneisses, limestones, shales, basalts and other volcanics, etc. An inspection of the granite outline shows that there are approximately 2400 miles of granite contact where the batholith occurs in Idaho. This includes the perimeter of the main intrusion, large inside patches of invaded rocks, and isolated islands. From a prospecting standpoint, this offers tremendous possibilities. There are many hundreds of miles still to be properly investigated. In fact, the whole outline may well be reprospected.

Prospecting should be carried on within, say, three miles of the granite contact both in the invaded rock and the granite. Where no great depth to underlying but unexposed granite is suspected, this distance could be indefinitely extended (various parts of Lemhi County, for example). The most likely areas will be those within a mile. But as a few deposits occasionally extend to the three mile limit, it is suggested.

DESCRIPTION OF COUNTIES

A discussion of the counties will include production figures and geological maps. There are many additional maps and reports on the State that are not included in this paper. Only those which, in the writer's opinion, were relative to gold have been selected.

As has been previously suggested, the writer believes that many of the very old discoveries have been forgotten or abandoned with the pass-

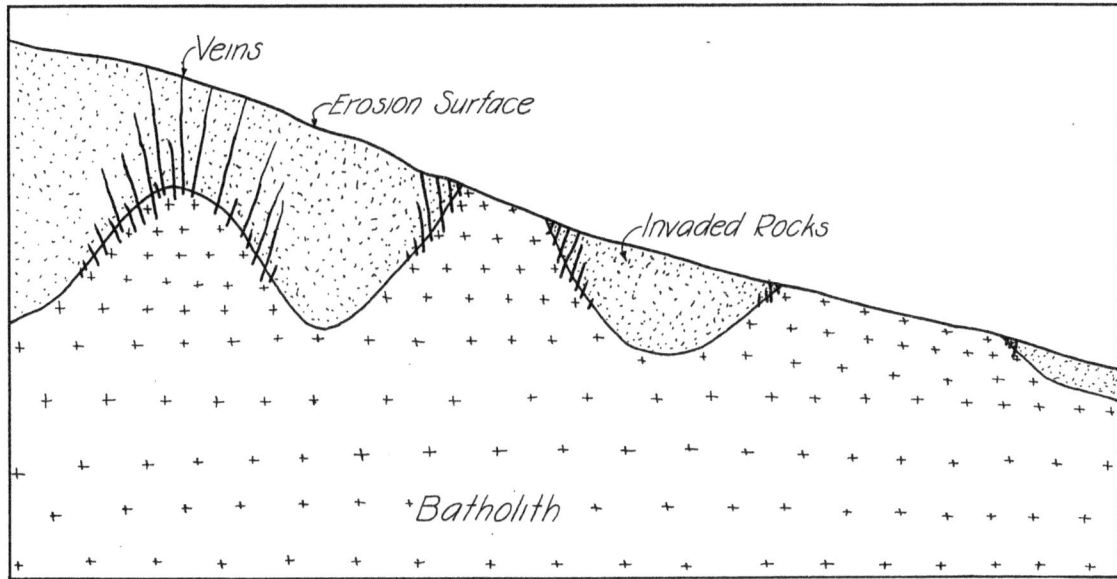

Figure 4.

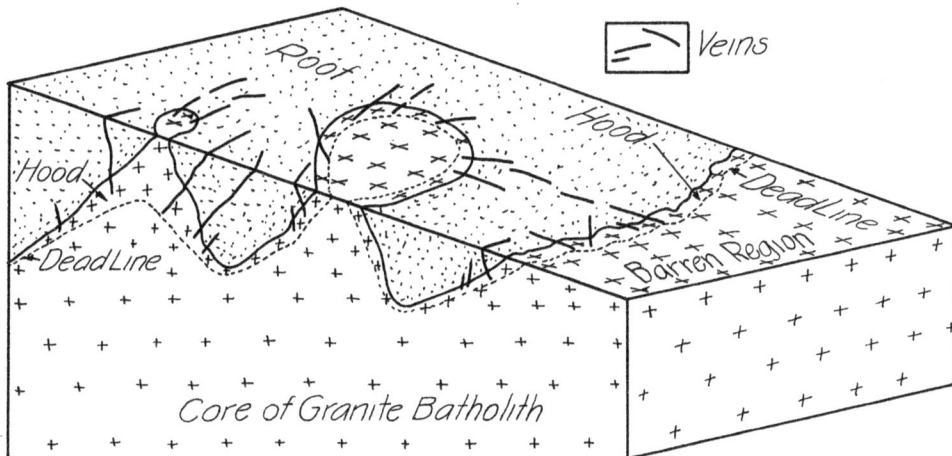

Figure 5.

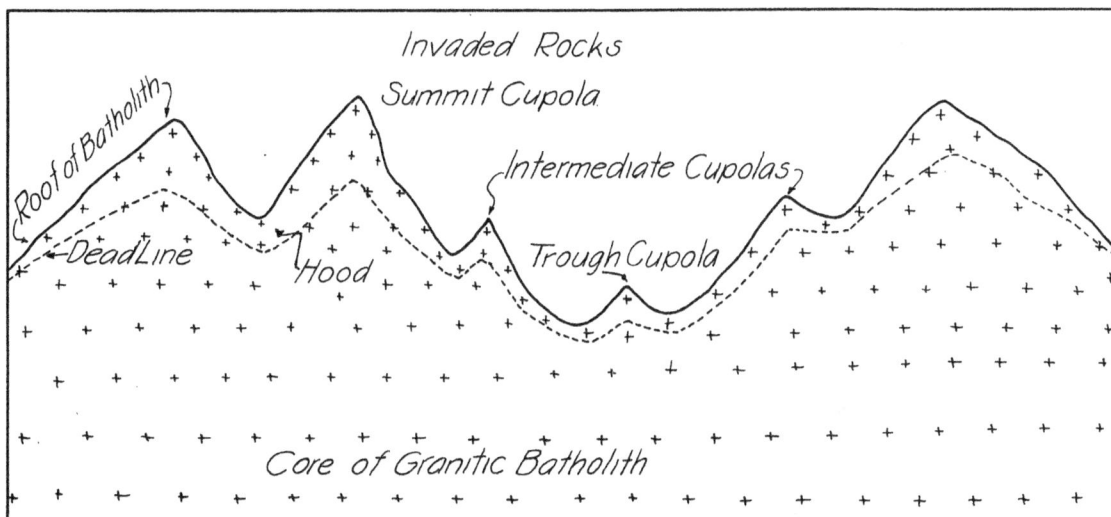

Figure 6.

age of time. Under present, and it is to be hoped still better future transportation facilities, and more economic mining and milling methods, many of these old workings should be reinvestigated. This is the excuse, such as it may be, for favoring the older publications. Historians, newspapers, and state executives of the past have painted a very rosy picture of mining during their day. Nevertheless, the old adage "where there is lots of smoke, there is bound to be a little fire" may very aptly be remembered when discounting their enthusiastic assertions.

Constant use should be made of Plate 1, as no consistent attempt will be made to call attention to changes in county boundaries or names. The present (1944) naming of the counties will be followed. Plate 2 also should be studied in connection with county geology. From the publication date of the reference quoted, the former boundaries, location, and county names can be established. Figure 1, the Index map, will act as a guide toward showing the various areas covered by the geological maps. All production is given in ounces except where otherwise noted.

Figure 7 (22) gives a general idea of the location of veins and mineralized areas as known many years ago. Metals other than gold are shown on the map. This figure includes a certain amount of information not appearing on other illustrations given in this report.

In the production tables accompanying each county, the figures in parentheses are taken from the Idaho State Mine Inspectors Annual Reports; the source of his information is not known. Placer data was not available before 1905. Output by counties was uncertain, even if known, before 1880. (Boise and Idaho Counties are possible exceptions). For this reason the state total does not check the county total. The totals given cover only the period shown by the tables, and it must be remembered that there was production before this time in the older counties. There must have been a considerable production for 1860-61 and 62. No conclusive record of it was found. The earliest data given by Hailey (23) are for 1863 for which he gives $5,000,000. See former excerpts from Bancroft.

ADA

Total		Total	Placer		Total	Placer
1880— 304		1902— 715			1923— 42 (13)	14
1881— 250		1903— 516			1924— 31 (58)	23
1883— 250		1904— 84			1925— 33 (53)	5
1884— 250		1905— 104	104		1926— 6 (10)	6
1885— 143		1906—1,278 (51)	60		1927— 83 (22)	2
1886— 102		1907— 370	101		1928— 17	5
1887— 113		1908— 289 (334)	69		1929— 23 (18)	
1888— 113		1909— 167 (534)	126		1930— 22	22
1889— 160		1910— 69 (115)	16		1931— 51 (34)	32
1890— 115		1911— 142 (142)	106		1932— 97 (65)	94
1891— 245		1912— 487 (370)	68		1933—106 (186)	102
1892— 272		1913— 138 (420)	98		1934—170 (8)	135
1893— 363		1914— 126 (210)	123		1935—170	145
1894— 739		1915— 146 (700)	93		1936—655	644
1895—1,966		1916— 129 (550)	129		1937—521	512
1896—1,323		1917— 282 (575)	267		1938—577	400
1897— 938		1918— 58 (344)	47		1939—621	594
1898— 670		1919— 12 (17)	12		1940— 74	64
1899— 879		1920— 3 (150)	3		1941—453	447
1900—1,621		1921— 21 (125)	21		1942—516	496
1901—1,032		1922— 60 (14)	16			

Total production, 21,312 oz.; placer, 5,201 oz.

Geology.—Figure 10, the Boise Quadrangle, includes a portion of Ada County. For a discussion of this map see Boise County.

ADAMS

	Total	Placer			Placer			Total	Placer
1911—	52 (52)	36		1922— 0 (0)			1933—	1	1
1912—	16 (85)	5		1923— 0 (195)			1934—	0	
1913—	21 (250)	10		1924—22 (94)	22		1935—1,292		1
1914—207 (850)		19		1925— 7 (226)	7		1936— 887		
1915—244 (750)		20		1926— 1 (2)	1		1937—3,984		
1916—152 (350)				1927— 1 (7)			1938— 293		19
1917— 0 (200)				1928— 3			1939— 543		36
1918— 18 (29)		18		1929— 5			1940— 331		28
1919— 1				1930— 0			1941— 257		16
1920— 0 (141)				1931— 3			1942— 16		
1921— 0 (2)				1932— 6					

Total production, 8,363 oz.; placer, 239 oz.

Geology.—There are reports of former placer operations in the vicinity of Bear. The bulk of the gold is probably from the copper ore shipped from time to time. Possibly some Snake River placering.

BANNOCK

	Total		Total		Total	Placer
1895—458		1902—631		1909— 0 (15)		
1897—453		1903—511		1910— 0 (83)		
1898—514		1904— 60		1920—12		
1899—768		1905— 73		1925— 2		2
1900—533		1906— 2		1941— 1		
1901—182		1908— 0 (10)				

Total production, 4,200 oz. This is no doubt all placer and was probably from the Snake River.

Geology.—For remarks concerning gold deposits see section on Snake River Placers.

BEAR LAKE

Total	Total	Total
1907—9	1924—0 (5)	1927—0 (40)
	1925—0 (5)	

Total, 9 oz. There was a fraction of an ounce for a few other years. There has been a little copper and a small amount of lead-silver ore shipped from Bear Lake County. Possibly the gold came from these sources.

Geology.—No geology on gold deposits published. County consists of sedimentary rocks; limestones, shales, etc. No known intrusive igneous rocks. The occurrence of the copper deposits west of Paris and east of Franklin (in Franklin County) are not satisfactorily explained. There may be a granitic intrusion at no great depth.

BENEWAH

Total	Total	Total
1915—235	1931— 1	1937—11
1916—392	1932—41 (8)	1938— 4
1917—154	1933—16 (13)	1939— 3
1918— 34	1934—29	1940—23
1919—117 (116)	1935—26	1941— 2
1920— 24	1936—16	1942—14
1925— 2		

Total production, 1,140 oz.

Geology.—Figure 1 shows that there was no map for this area. Referring to Plate 2, we find a little granite outcropping in the southeastern part of the country. There may be other contacts which have not been mapped.

Clearwater

Grangeville

Mt. Idaho

American (South Fk Clearwater)

G Elk City

IDAHO

MONTANA

G Buffalo Hump

Florence G

River

Salmon

River

G Gibbonsville

Warren

G

G Shoup

Salmon

River

L E M H I

G Leesburg

Snake

River

Cu   Cu G

Seven Devils

Cu

South Fk

Middle Fk

G Yellow Jacket

Cu
SL

WASHINGTON

Mineral

S

G Gold Fork

North Fork

Weiser

R

Weiser

Weiser

B O I S E

SL Sheep Mountain

Seafoam GS

GS Custer

SL Challis

C U S T E R

Salmon

G Deadwood

South Fk

G Stanley Basin

Pauette

C A N Y O N

G

Placerville

Willow Creek

G G Pioneerville

Idaho City

Banner S

SL Germania

Atlanta
GS

G
S

S

Vienna

GS

B L A I N E

G
G

BOISE

Boise

R

Neal

G
G

G

G

Rocky Bar
GS

SL

SL SL

G

G

Hailey

Caldwell

O R E G O N

A D A

E L M O R E

Mountain Home

SL SL

G

Camas

G

Big Little

Wood

Wood

R

R

Shoshone

GS || GS
Silver City

S

Flint

South Mountain
SL

Snake

Bruneau

R

O W Y H E E

LINCOLN

River

Salmon Cr

C A S S I A

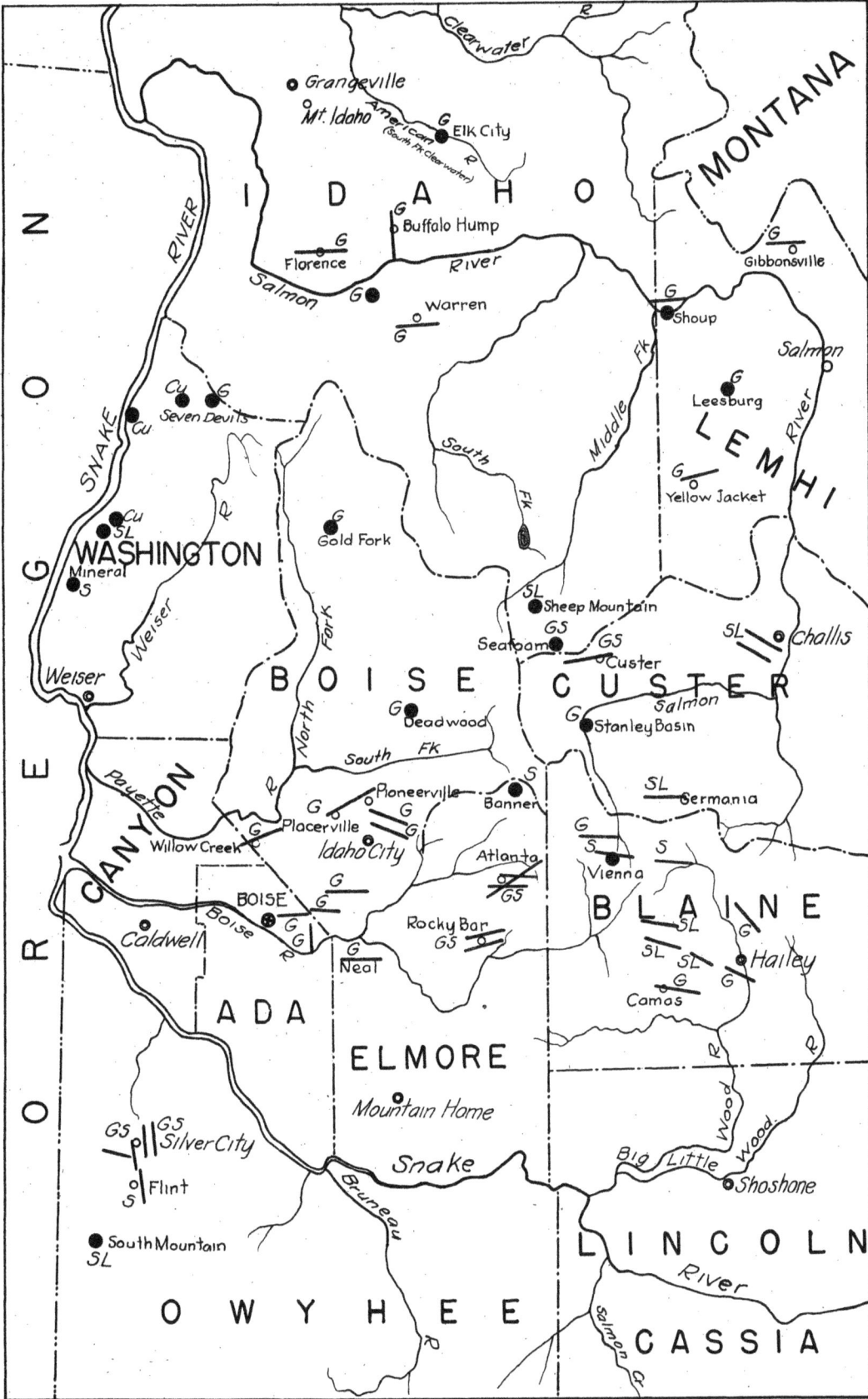

25   0   25   50   75 MILES

/ ● Veins and mineralized areas.   G-Gold; S-Silver; L-Lead; Cu-Copper.

Fig. 7—Map of western central Idaho

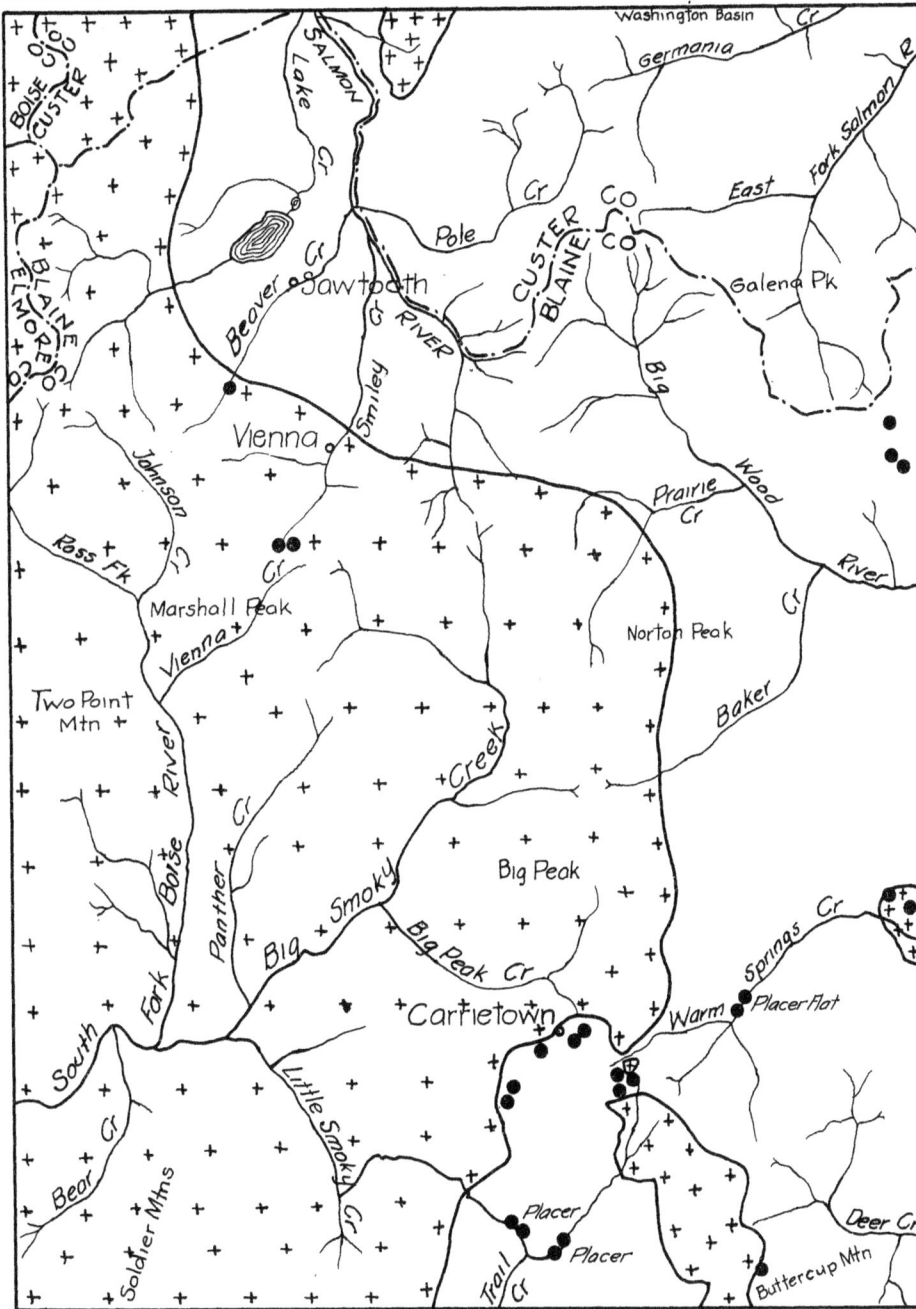

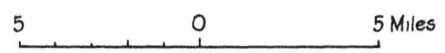

Fig. 8—Geologic sketch map of Sawtooth Quadrangle

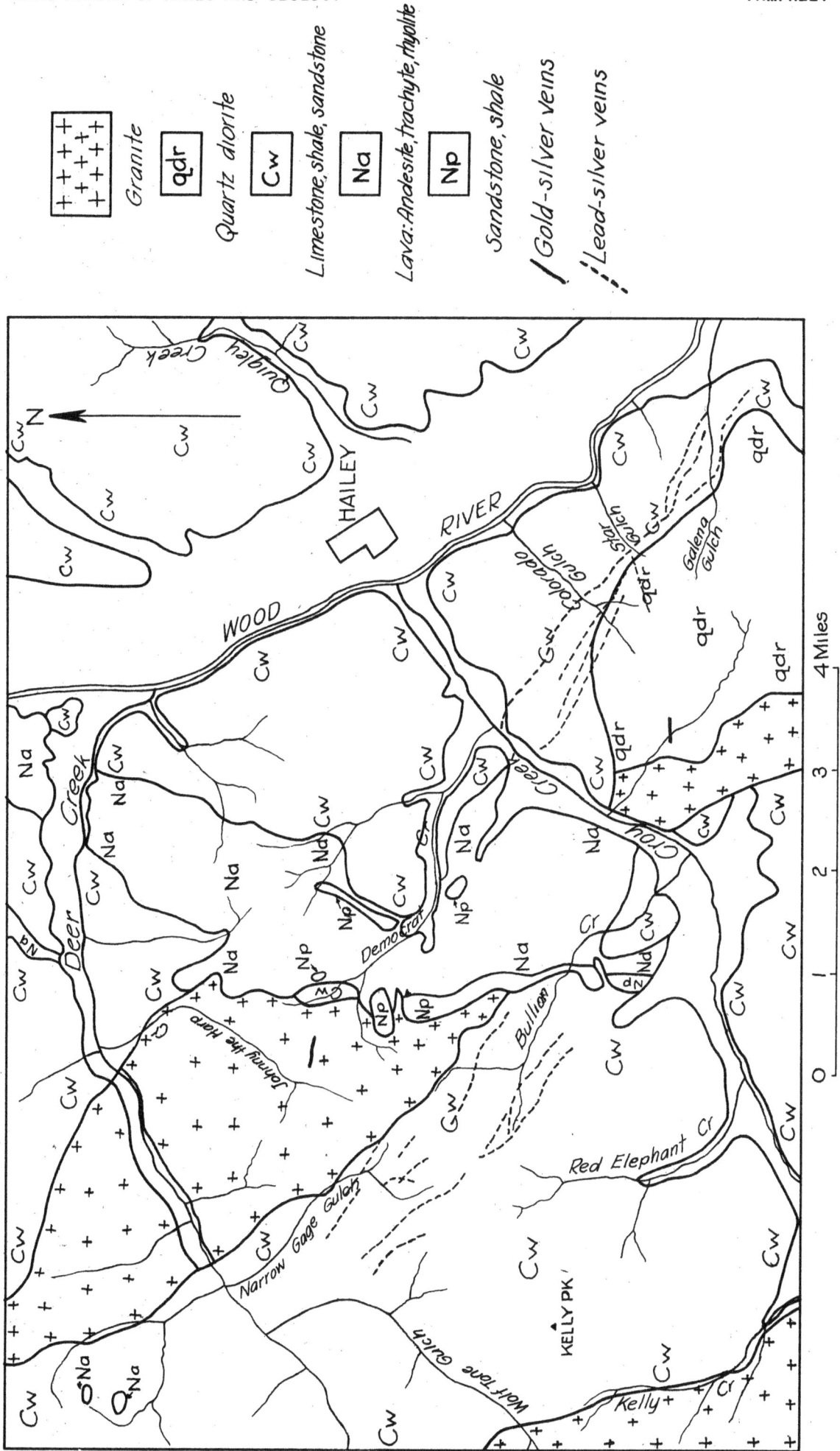

Fig. 9—Geology of Mineral Hill mining district, Hailey

BINGHAM

Total		Total	Placer		Total	Placer
1885—3,626		1901—639			1916— 2	2
1886—3,442		1902—579			1917—13	13
1888—1,500		1903—641			1918— 2	2
1889—1,000		1904—498			1921—24 (43)	24
1890— 377		1905—276	276		1922— 1 (1)	1
1891— 572		1906—262 (73)	255		1923— 9 (16)	9
1892— 436		1907—377	375		1924—11	11
1893—1,356		1908—297 (238)	293		1925—35	35
1894—1,788		1909—172 (5)	156		1926— 1	1
1895—1,809		1910— 65 (48)	67		1931—10	10
1896— 756		1911— 80 (80)	80		1932— 7 (26)	7
1897— 747		1912— 72 (72)	72		1933— 6 (1)	6
1898— 711		1913— 58 (58)	58		1934—10	10
1899— 941		1914— 44 (44)	44		1935— 2	2
1900— 881		1915— 37 (37)	37		1936—70	2

Total production, 24,242 oz.; placer, 1,848 oz. A few slight differences will be noted for several years. Undoubtedly the total production should be credited to Snake River placers. There is no place else in the county to mine gold. Sometime previous to 1911, Bingham County included most of the Snake River in this region until it left Idaho at the Wyoming line. However, Bancroft (see previous comments) calls attention to other placering operations when this was part of Oneida County.

Geology.—See Snake River placers.

BLAINE

(Formerly Alturas)

Total		Total	Placer		Total	Placer
1874— 7,500		1901— 939			1922— 853 (856)	
1880— 7,454		1902— 725			1923— 487 (459)	
1881— 8,500		1903—1,143			1924— 147 (150)	
1882— 8,750		1904— 191			1925— 153 (154)	
1883— 7,500		1905— 346	258		1926— 50 (45)	
1884— 5,000		1906— 658 (885)	110		1927— 68 (102)	
1885— 4,129		1907—1,311	3		1928— 164 (223)	
1886—10,338		1908— 127 (413)	63		1929— 202 (327)	
1887—18,128		1909—1,955 (1,856)	6		1930— 163 (117)	
1888—15,000		1910—2,048 (600)			1931— 29 (27)	
1889— 1,382		1911—1,044 (1,044)	12		1932— 101 (109)	
1890— 3,434		1912— 901 (750)	5		1933— 278 (258)	
1891— 2,061		1913— 222 (565)	2		1934— 81 (221)	
1892— 1,102		1914— 678 (851)	7		1935— 245	2
1893— 824		1915—1,348 (1,405)	3		1936— 489	1
1894— 836		1916— 899 (1,500)			1937— 1,413	
1895— 3,159		1917— 506 (600)			1938— 5,056	8
1896— 3,236		1918— 634 (898)	3		1939—13,539	3
1897— 557		1919— 640 (630)			1940—10,216	5
1898— 537		1920— 635 (625)			1941— 6,783	2
1899— 1,504		1921—1,139 (1,151)			1942— 5,059	
1900— 1,666						

Total production, 176,262 oz.; placer, 492 oz.

Geology.—From the Index map, location 7 and 8 are found to contain maps of Blaine County. Also Fig. 3 gives the early day location of mines.

Figure 8 was taken from Umpleby (24) with some additions from Ross (25).

The production has been mostly from lead-silver ores. Large low grade quartz-gold veins are not uncommon.

According to Ross (26) the ore deposits are largely replacement, are formed in shear zones, and are genetically related to the Idaho batholith. Most of the veins are in sedimentary rock located close to the granite contact. The characteristic ore minerals are galena, sphalerite, pyrite, and tetrahedrite in a gangue of quartz, siderite, and altered country rock.

Figure 9 gives the geology in the vicinity of Hailey. This district is noted for its lead-silver output. Some gold had been found in veins in the quartz diorite in the southwestern part of the sheet and in the granite area in the central and northwestern part.

The map is taken from Lindgren's report on the district (27). Occasionally appreciable quantities of gold were found in the lead-silver veins. In such cases the veins also contained quartz (28). About 12 miles southwest of Hailey on the Fairfield road is the "Hailey gold belt." The combination here is quartz, calcite, siderite, pyrrhotite, pyrite, and chalcopyrite. There are numerous

quartz vein systems in granite. Recent investigation suggests this area as a possible gold producer*.

In the quartz diorite area the vein matter is altered granite filled with small quartz seams. Occasionally some siderite, and considerable massive pyrite, pyrrhotite, and chalcopyrite with lesser amounts of galena, sphalerite, and arsenopyrite. The gold values are said to occur in the chalcopyrite and pyrite.

* Report in preparation by Idaho Bureau of Mines and Geology.

BOISE

Total	Total	Placer	Total	Placer
1863—150,000	1890—18,695		1917—10,521 (10,500)	1,855
1864—200,000	1891—17,253		1918—10,791 (10,810)	2,493
1865—250,000	1892—18,209		1919—20,243 (14,278)	271
1866—250,000	1893—13,547		1920—12,858 (8,681)	77
1867—215,000	1894—15,859		1921—12,389 (13,817)	514
1868—215,000	1895—16,429		1922— 7,197 (6,255)	406
1869—150,000	1896—15,770		1923— 4,583 (4,537)	503
1870—135,000	1897—11,274		1924— 3,703 (3,640)	123
1871—100,000	1898—10,008		1925— 2,249 (2,250)	563
1872— 50,000	1899—19,055		1926— 3,956 (3,702)	1,799
1873— 40,000	1900—17,835		1927— 7,366 (7,389)	2,766
1874— 35,000	1901—18,104		1928—11,291 (11,079)	2,075
1875— 30,000	1902—16,481		1929—12,566 (12,925)	2,568
1876— 30,000	1903—14,903		1930—13,260 (13,007)	2,778
1877— 25,000	1904—14,265		1931— 9,992 (10,234)	1,092
1878— 25,000	1905— 7,360	6,351	1932— 2,636 (2,701)	1,365
1879— 20,000	1906—11,709 (13,113)	6,661	1933— 3,236 (2,025)	1,728
1880— 38,254	1907—11,382	8,040	1934— 9,365 (10,761)	5,070
1881— 15,000	1908— 9,642 (10,451)	5,699	1935—12,976	5,044
1882— 14,500	1909—10,451 (11,673)	6,941	1936—21,675	12,386
1883— 28,250	1910— 9,647 (8,505)	6,273	1937—26,878	19,763
1884— 20,000	1911—17,099 (17,099)	12,857	1938—27,730	21,351
1885— 30,980	1912—24,248 (25,071)	21,757	1939—21,973	18,715
1886— 18,910	1913—29,996 (29,000)	25,634	1940—27,197	23,065
1887— 21,656	1914—29,085	22,867	1941—23,438	20,882
1888— 14,150	1915—26,854 (29,810)	18,540	1942—14,731	13,134
1889— 13,280	1916—22,739 (19,318)	14,279		

Total production, 2,917,679 oz.; placer, 318,285 oz. The period 1863-1872 was more than likely mostly placer. From 1862 to 1863 a substantially unknown quantity was produced.

Geology.—Area 11, showing southwestern Boise County and northeastern Ada County, is from the Boise Quadrangle sheet (29). The remarks concerning the geology of this area are taken from the folio and also from Lindgren's report on the area (30). (See Fig. 10 (29).

The Boise Basin (31) is represented by area 10 and Fig. 11 (31).

The general geological features are best told by quoting directly from Lindgren's (31) report:

"Throughout the Boise Ridge and the Idaho Basin the primary gold deposits present a certain similarity. They are all contained in granitic rocks

or associated dikes. They are all either fissure veins or impregnations connected with fissures. Nearly all of the fissures have a direction ranging from east-west to northeast-southwest, the chief exception to this rule being found in the Black Hornet district. The dip is ordinarily to the south at angles of from 45 deg. to 89 deg., but in the Willow Creek and Rock Creek districts similar dips to the north are found. The prevailing direction of the fissures is the same as that of an often well-developed system of joint planes commonly seen in the Boise Ridge. Finally, the ores are, on the whole, of a similar character, consisting chiefly of auriferous pyrite, arsenopyrite, zinc-blende, and galena in a gangue of quartz or, more rarely, calcite. The fresh ores from deeper levels contain a variable percentage of free gold. Rarely, however, is more than 60 per cent of the total value caught on the amalgamating plates. Gold predominates largely in the value of the ore, though sel-

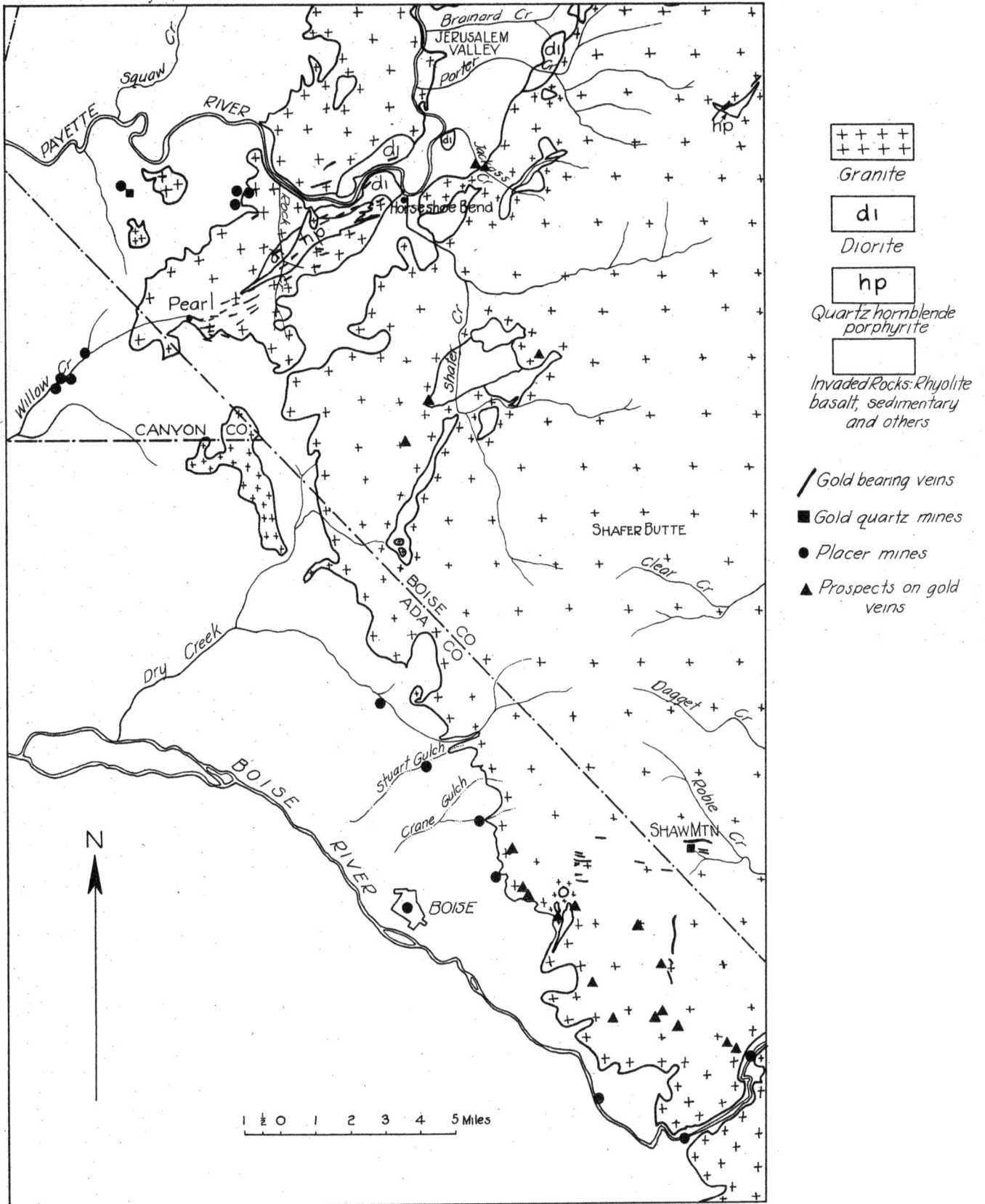

Fig.10—Boise Quadrangle

Legend:

Granite

di
Diorite

hp
Quartz hornblende porphyrite

Invaded Rocks: Rhyolite basalt, sedimentary and others

Gold bearing veins
Gold quartz mines
Placer mines
Prospects on gold veins

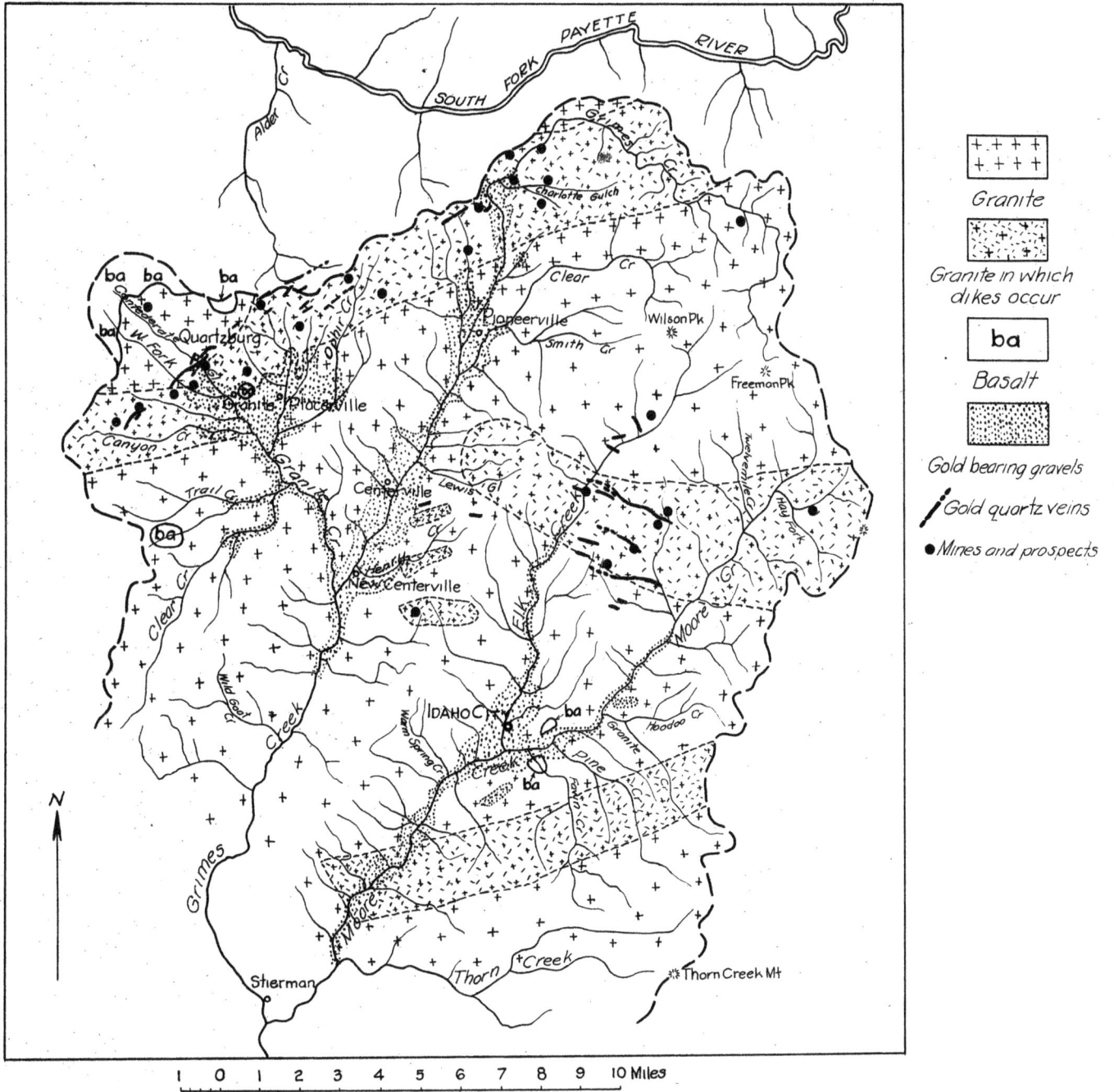

Fig.II—Geologic map of the Boise Basin, Boise County

dom by weight, for in the ordinary ores the weight of the silver considerably exceeds that of the gold. The alteration of the country rock in the vicinity of the vein is throughout of the same character."

An inspection of Fig. 10 shows the location of mines, prospects, and placers at the time of its original publication. The close association of the veins with the granite contact should be noted.

An examination of Fig. 11 suggests some interesting possibilities. This map is one including the features of both Lindgren's and Ballard's. Many of the veins and prospects and placer gravel areas on Lindgren's map were not given on Ballard's. The present figure contains these.

The map includes only the area in the Basin.

The veins and prospects occur almost without exception in the rock indicated as "granite in which dikes occur." Large areas of this formation appear to be unexplored; and the extension of the same formation to the east beyond the confines of the Basin offer interesting possibilities. An investigation for placer deposits to the east is well worth consideration.

It is granted that, so far as the Basin proper is concerned, the bulk of the placer ground has been mined; not once but several times over. There are, however, many streams and it is entirely possible that patches of good ground have been overlooked.

Figure 12 shows a few typical examples of the auriferous gravel deposits and of the gold bearing quartz veins (32).

BONNER

	Total	Placer		Total				Total	Placer
1907—	2		1919—	1	(19)		1931—	0	
1908—	8	(125)	1920—	1			1932—	3 (8)	0
1909—	25	(361)	1921—	8	(7)		1933—	7 (24)	
1910—	41	(100)	1922—	1,642	(1,864)		1934—	6	0
1911—	9	(9)	2	1923—	2,604	(2,591)	1935—	17	
1912—	27	(329)	1924—	1,130	(1,130)		1936—	46	
1913—	60	(370)	5	1925—	693	(625)	1937—	257	
1914—	33	(310)	1926—	168	(151)		1938—	65	
1915—	34	(34)	9	1927—	3	(3)	1939—	38	
1916—	152	(152)	1928—	2	(5)		1940—	10	
1917—	1	(100)	1929—	6	(5)		1941—	3	
1918—	0	(70)	1930—	1	(1)		1942—	3	

Total production, 7,106 oz.; placer, 16 oz.

Geology.—No satisfactory map, so far as gold is concerned is available. Anderson (33) has published a report on the Clark Fork district.

BONNEVILLE

	Total	Placer		Total	Placer		Total	Placer
1911—	138 (138)	138	1922—	5	5	1933—	60	58
1912—	244 (242)	235	1923—	40 (40)	40	1934—	52	50
1913—	220 (220)	220	1924—	15	15	1935—	128	128
1914—	191 (191)	191	1925—	18 (15)	18	1936—	142	142
1915—	219 (219)	218	1926—	39 (11)	37	1937—	144	118
1916—	255 (227)	255	1927—	20 (21)	20	1938—	124	115
1917—	106 (150)	105	1928—	24	24	1939—	92	92
1918—	108 (157)	108	1929—	10	10	1940—	77	77
1919—	113	113	1930—	12	12	1941—	78	77
1920—	17	17	1931—	19	19	1942—	33	33
1921—	111 (110)	111	1932—	63 (59)				

Total production, 2,917 oz.; 2,801 oz.
Geology.—See Snake River Placers.
Bancroft (34) remarks about early day production from the Mount Pisgah region.

There was copper production carrying gold, gold placers on McCoy Creek and other eastern slope creeks, and Snake River gold. At that time Oneida County covered the area.

BOUNDARY

Total			Total			Total	
1915—17 (17)			1923— 0 (31)	96,243*		1933—68	
1916— 9 (9)	47,517*		1924— 0	100,465*		1934— 0	
1917—32 (200)	117,000*		1925—268 (268)	59,812*		1935— 3	1†
1918— 0 (60)	129,644*		1926— 0	58,676*		1937— 1	1†
1919—33	107,312*		1927— 0	13,549*		1938—34	
1920—25	99,000*		1928— 3	19,744*		1939—67	
1921—29 (20)			1929— 0			1942—11	
1922— 2 (2)	60,316*		1932— 2 (55)	2†			

———

* Silver, ounces, from the State Mine Inspector's reports. Given to show that there is some metal production from this county.
† Placer.

Total production, 604 oz.; placer, 4 oz.; silver for period given, 909,278 oz.

Geology.—The gold production of Boundary County is relatively insignificant. On the other hand, a fair amount of lead and silver have been produced. The map for area 14, and included as Fig. 13, is from the Kirkham and Ellis report on Boundary County (35). The authors suggest that, because of glaciation, the greater part of the veins have been removed. It will be noted that the majority of the veins seem to be associated with the so called basic sills.

BUTTE

Total		Total		Total	
1917—85 (50)		1926— 2 (2)		1935— 1	
1918— 6 (49)		1927— 7 (3)		1937— 48	
1919— 3		1928— 1 (2)		1938—141	
1920— 0		1929— 4 (21)		1939—245	
1922— 2 (2)		1930— 3 (3)		1941— 5	
1924— 2 (2)		1932— 0		1942— 2	
1925— 1 (1)		1934—13			

Total production, 571 oz.; placer, a trace.

Geology.—Butte County, at least to date, has shown no signs of becoming a gold producer. An inspection of Plate 2 shows no granite except in the extreme northeastern part of the county.

CAMAS

Total	Placer	Total	Placer	Total	Placer
1917—396 (500)		1926— 21 (27)	1	1935—1,514	1,190
1918—325 (245)		1927— 1 (1)		1936— 740	456
1919— 3		1928— 30 (30)		1937— 194	14
1920— 22		1929— 41		1938— 442	241
1921— 54 (26)	3	1930— 47 (54)		1939— 304	16
1922—232 (224)	8	1931—165 (138)	11	1940—2,152	17
1923— 20 (20)		1932— 34 (60)	2	1941— 618	17
1924— 72 (59)	8	1933—180 (347)	8	1942— 265	2
1925— 26 (16)		1934—387 (192)	55		

Total production, 8,249 oz.; placer, 2,049 oz.

Geology.—See Blaine County and Fig. 8. Until rather recently, Camas County was a part of Blaine County. An inspection of a highway map shows an almost total lack of roads and towns in northern Camas County (the county is fairly accessible by Forest Service roads). Nevertheless, during the eighties there was a very considerable mining activity. On the strength of this early work the area should be reprospected.

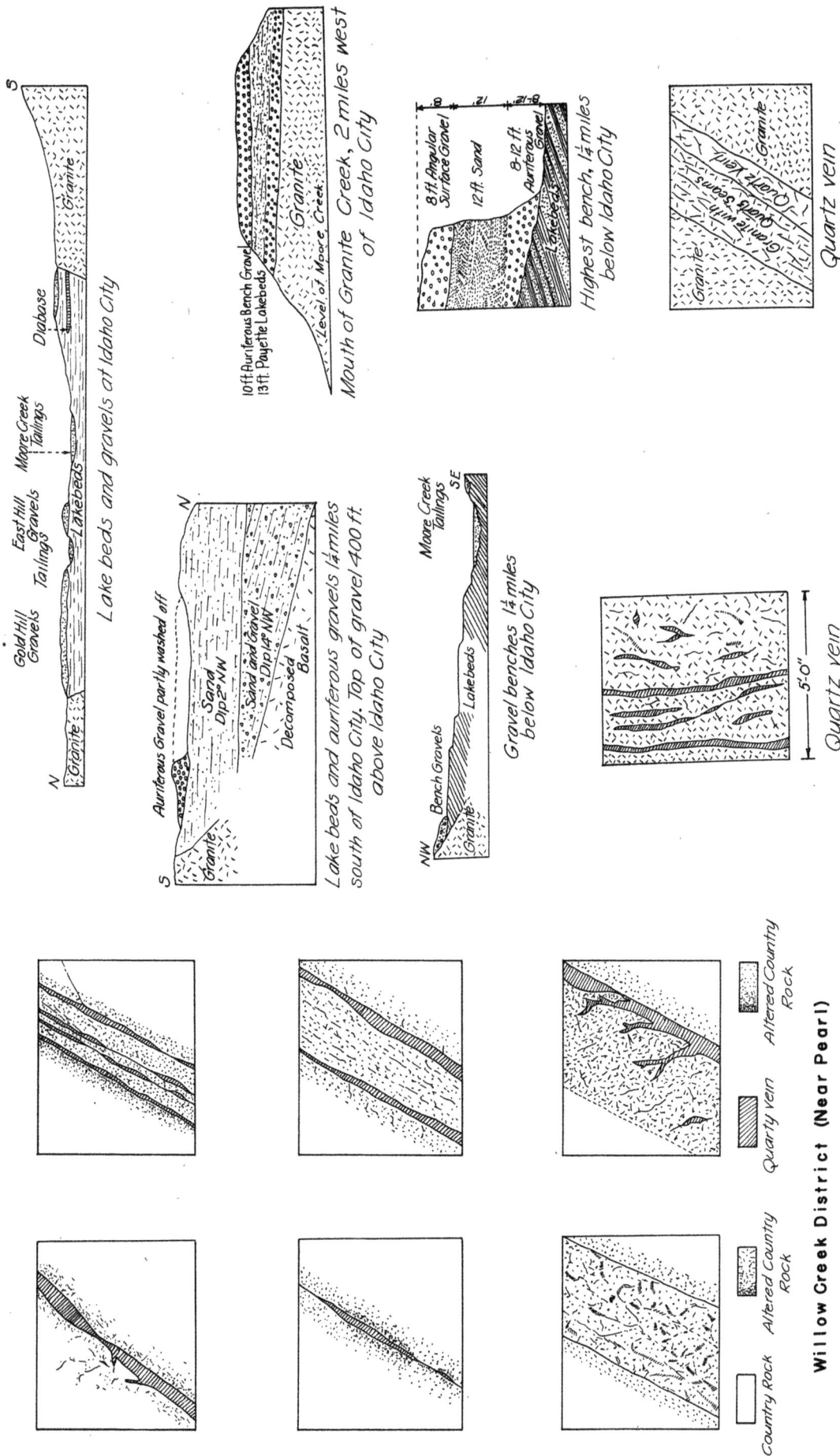

Fig.12—Typical auriferous gravel deposits and quartz veins

Boise Basin

Lake beds and gravels at Idaho City

Mouth of Granite Creek, 2 miles west of Idaho City

Highest bench, 1¼ miles below Idaho City

Lake beds and auriferous gravels 1¼ miles south of Idaho City. Top of gravel 400 ft. above Idaho City

Gravel benches 1¼ miles below Idaho City

Quartz vein

Quartz vein

Willow Creek District (Near Pearl)

Country Rock

Altered Country Rock

Quartz vein

Altered Country Rock

MONTANA

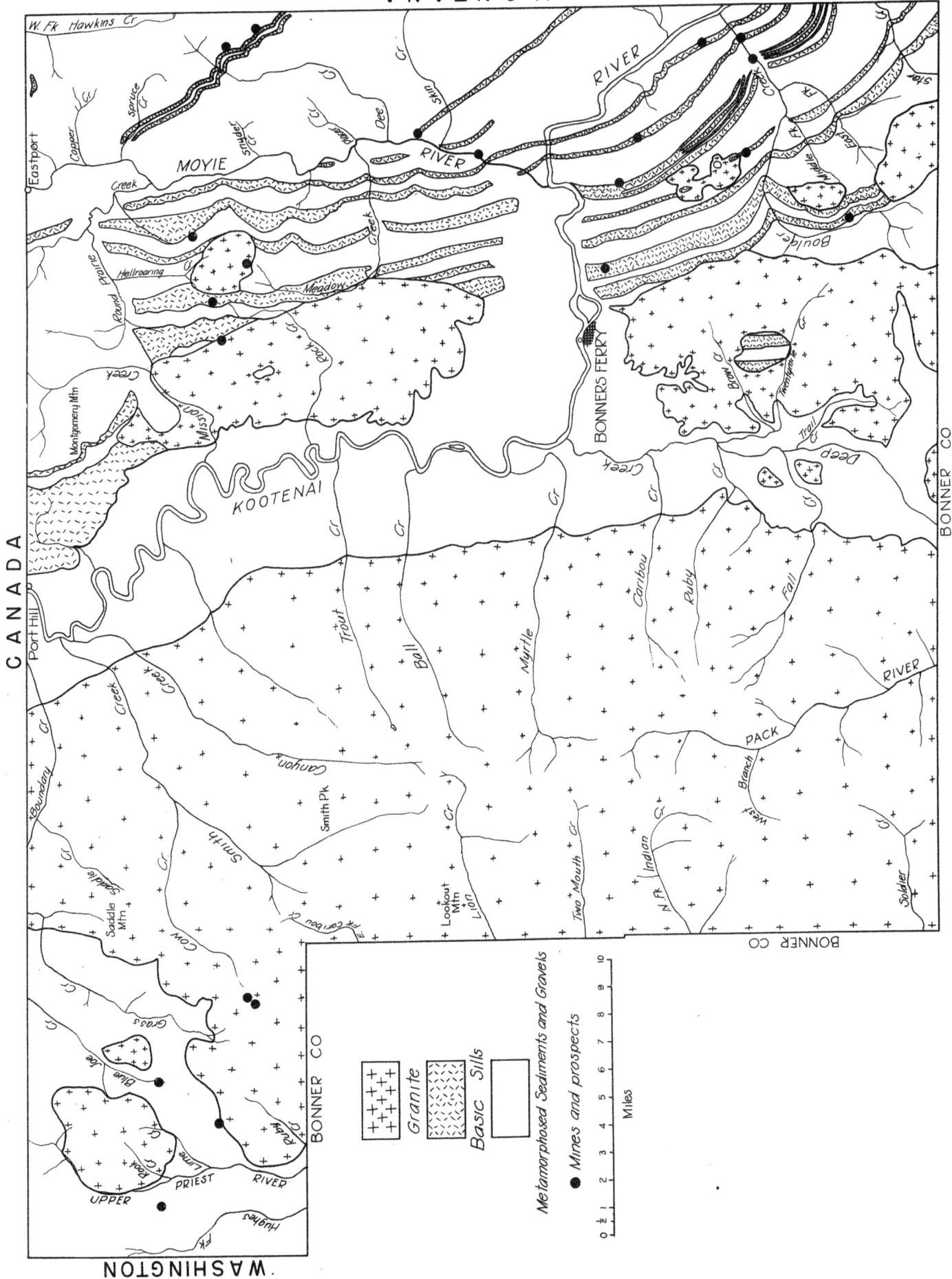

Fig.13—Geologic map of Boundary County

Granite

Basic Sills

Metamorphosed Sediments and Gravels

● Mines and prospects

Miles

CANYON

Total	Placer		Total		Placer		Total		Placer
1893— 323			1907—15		14		1925— 1		1
1894—1,029			1908—38 (46)		38		1926— 1		1
1895— 823			1909—15 (39)		15		1927— 2		2
1896— 522			1910— 4 (4)		4		1932— 7 (30)		7
1897— 422			1911—10 (10)		10		1933— 17 (10)		17
1898— 413			1912—12				1934— 3		3
1899— 526			1913—25				1935— 3		3
1900— 465			1914—27				1936— 1		1
1901— 0			1915—18 (29)		18		1937— 3		3
1902— 255			1916—83 (67)		83		1938— 3		3
1903— 443			1920— 0 (74)				1939—642		642
1904— 8			1921— 2 (4)		2		1940— 7		7
1905— 71	71		1922— 9				1941— 3		3
1906— 52	52		1923— 3 (3)		3		1942— 0		0

Total production, 6,306 oz.; placer, 1,003 oz.

Geology.—See Snake River Placers.

An inspection of the production table shows that from 1905 on the total production and the placer production figures are identical. This gold could come from both the Boise and Snake Rivers. It is impossible to assign a definite amount to either source.

CARIBOU

No gold reported from here.

CASSIA

Total		Total		Placer		Total		Placer
1880—1,542		1898—1,031				1915—44		
1881—1,250		1899— 875				1916—55		
1882—1,250		1900—1,935				1917—31		
1883— 500		1901—1,758				1918—23		
1884— 250		1902—1,890				1919—79		
1885— 646		1903— 545				1920— 0 (2)		
1886— 288		1904— 281				1924—12 (12)		
1888— 225		1905— 141		140		1925—25 (25)		
1889— 600		1906— 163		161		1926—27		
1891—1,200		1907— 61		42		1931— 2		2
1892— 714		1908— 24 (15)				1934— 4		3
1893— 510		1909— 27 (2)		27		1936— 1		
1894—1,108		1910— 63 (65)				1937—11		
1895— 980		1911— 19 (19)		4		1939— 2		
1896— 896		1913— 23				1940—19		
1897—1,057		1914— 1				1941—11		

Total production, 22,209 oz.; placer, 379 oz.

Geology.—See Snake River Placers.

There is no map, so far as gold is concerned, on Cassia County. The bulk of the gold production as shown in the table came from the Snake River.

Very minor lead-silver and a little copper production produced a small amount of gold. Probably 99 per cent, at least, is Snake River. For the geology and ore deposits of the county see Anderson's report (36).

CLARK

Total		Total			Total	
1922—0		1925—0 (0.2)			1928—0	
1923—2 (2)		1926—0.1 (0.1)			1929—0 (248)	
1924—0		1927—0			1930—1	

Total production, 3 oz.; placer, 0.

Geology.—A very brief, and not too informative report covering Clark County has been made (37).

Northwest of Kilgore (38)*, and a few miles south of the State line, there are indications of both vein and placer gold. The country rock appears to be andesite and rhyolite, in many places highly brecciated and recemented.

---

* Bancroft mentions gold production in the eighties from a locality which appears to be near here.

CLEARWATER

Total	Placer	Total	Placer	Total	Placer
1911—1,972 (1,972)	1,972	1922—293 (293)	293	1933—1,404 (598)	1,371
1912—1,948 (2,692)	1,939	1923—296 (296)	296	1934—1,467 (556)	1,444
1913— 541 (2,082)	526	1924— 42 (19)	42	1935—2,275	2,273
1914— 911 (2,184)	904	1925— 59 (41)	59	1936—1,629	1,629
1915—1,363 (2,318)	1,357	1926— 40 (40)	40	1937—2,199	2,175
1916—1,271 (1,700)	1,263	1927— 60 (27)	60	1938—1,101	320
1917— 839 (1,100)	839	1928— 25 (29)	25	1939—2,101	1,744
1918— 234 (217)	219	1929— 51 (46)	51	1940—3,417	3,388
1919— 568 (118)	567	1930— 45 (42)	45	1941—1,968	1,934
1920— 357 (595)	357	1941— 96 (69)	79	1942— 22	22
1921— 172 (194)	172	1932—370 (325)	345		

Total production, 29,136 oz.; placer, 27,750 oz.

Geology.—The production from Clearwater County is almost entirely placer. Previous to 1911 the production would be split between Nez Perce, Shoshone, and possibly Idaho Counties. (See Fig. 16 and Idaho County.) Area 3 on the Index map covers this county. A number of quartz veins have been prospected in the general vicinity of Pierce. There is little known about much of the county, so that it deserves investigation (39).

CUSTER

Total	Total	Placer	Total		Placer
1881—25,000	1903— 6,113		1923— 1,608 (1,574)		55
1882—15,000	1904— 3,664		1924— 1,393 (1,467)		40
1883—10,000	1905— 1,632	947	1925— 2,195 (2,197)		13
1884—10,000	1906— 5,450 (5,951)	670	1926— 357 (367)		26
1885— 5,678	1907— 8,508	417	1927— 946 (931)		96
1886— 5,804	1908—13,669 (14,329)	228	1928— 800 (904)		20
1887—10,580	1909— 4,443 (4,930)	331	1929— 2,535 (2,569)		55
1888—10,000	1910— 1,115 (2,800)	264	1930— 935 (848)		8
1889—14,000	1911— 4,347 (4,347)	340	1931— 64 (64)		52
1890— 1,904	1912— 2,416 (3,224)	244	1932— 186 (117)		145
1891— 3,720	1913— 5,416 (5,800)	192	1933— 307 (85)		108
1892— 1,166	1914— 2,396 (1,562)	203	1934— 313 (170)		158
1893— 1,115	1915— 4,101 (4,100)	49	1935— 322		93
1894— 1,253	1916— 3,509 (2,938)	199	1936— 357		80
1896— 5,166	1917— 3,232 (3,250)	85	1937— 449		122
1897— 5,110	1918— 2,577 (2,758)	50	1938— 461		262
1898— 2,653	1919— 740 (350)	5	1939— 1,976		159
1899— 3,079	1920— 1,605 (1,125)	57	1940— 5,116		942
1900— 2,070	1921— 1,384 (1,263)	72	1941—12,803		7,595
1901— 897	1922— 2,157 (2,159)	52	1942— 6,190		4,793
1902— 668					

Total production, 252,879 oz.; placer, 19,226 oz.

Geology.—The gold producing area of Custer County, with the exception of that saved as a by-product from the copper mining at Mackay and possibly some with the lead-silver from Bay Horse, seems to be confined to the western part of the county and placers along the Salmon River. Fig. 14 (40), and also Fig. 3, covering area 6, gives essentially the available information.

From Thompson Creek east the deposits are almost entirely silver and lead-silver, although at Bay Horse there is one good copper prospect which no doubt carries gold, and several quartz veins are reported to be gold bearing. West of Thompson Creek the output has been essentially gold and gold-silver.

Umpleby mentions gold-copper ores in the Loon Creek area. This type of ore is an auriferous chalcopyrite in a siderite-quartz gangue. Pyrite and pyrrhotite are rare. In some veins gold occurs equally in the quartz. The siderite carries much less gold than either the chalcopyrite or the quartz. The values are rather high grade.

The gold-silver veins are found in the Yankee Fork district. In 1913 their total production was estimated at $12,000,000, 40 per cent of which was gold. The gold-silver veins are enclosed in the schists, quartzites, tuffs, latites, andesites, basalts, and rhyolites. Wall rock alteration adjacent to the veins has been so intense that it is difficult, if not impossible, to distinguish between the above rocks. Small patches of pyrite are abundant in the wall rock near the vein. The vein filling is mostly quartz, but may contain considerable calcite.

It is enlightening to quote Umpleby (41) regarding the ore shoots.

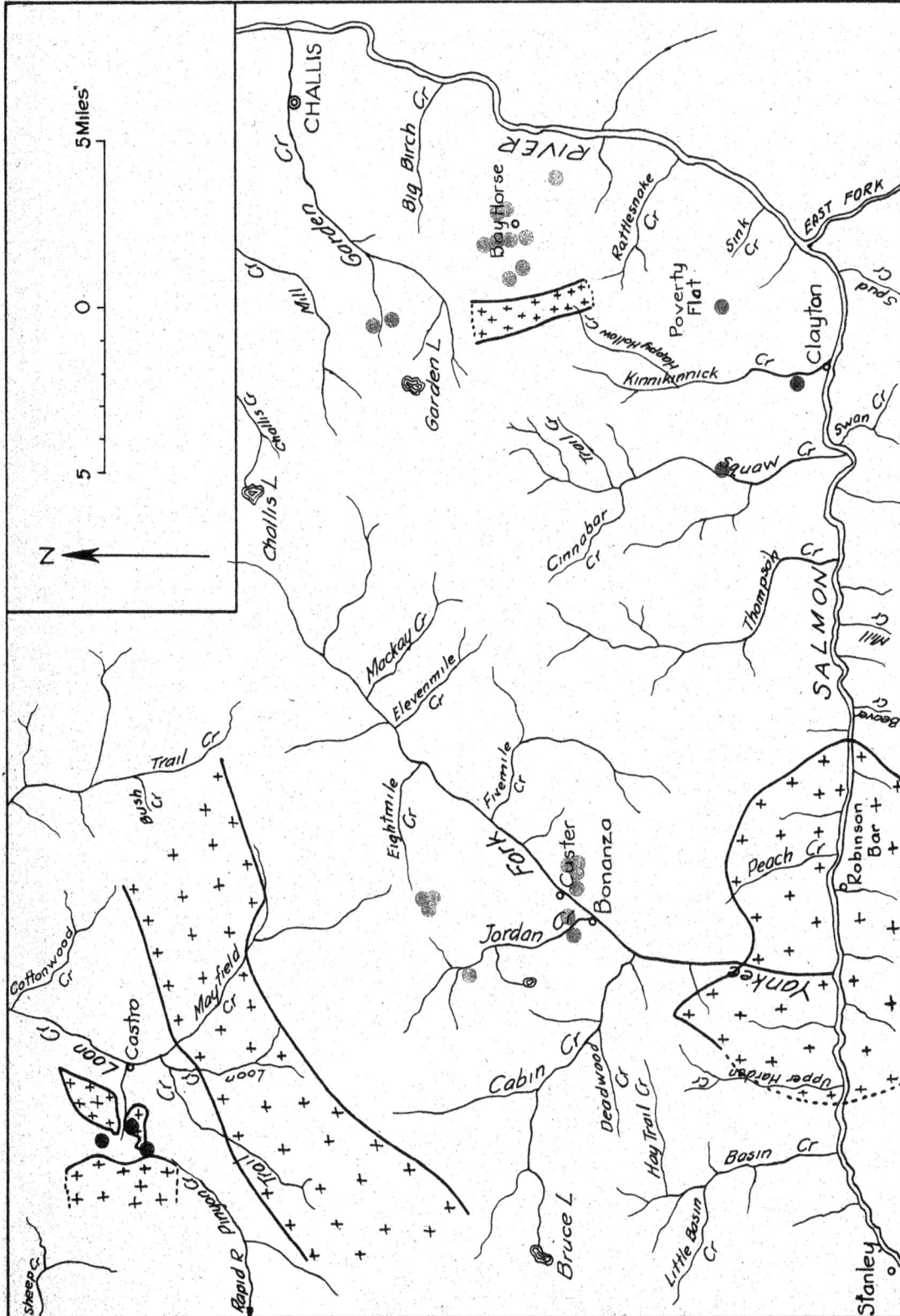

Fig.14-Geologic sketch map of a portion of northwestern Custer County

"The ore shoots of these deposits comprise (1) segments of otherwise comparatively barren quartz-calcite veins, (2) disseminations of stock-like form not related to fissure fillings, and (3) chimneys of ore apparently independent of fissure veins. The most important deposits are of the first type, which represents all the mines except the Golden Sunbeam and the Montana. The Golden Sunbeam was a great stock-like body of mineralized tuff, and the Montana was a chimney of ore about 10 by 20 feet in cross section."

The unaltered ore is a fine-grained quartz, containing calcite, chalcedony, opal, and adularia. Pyrite is abundant, the other sulfides only sparingly so. Tetrahedrite, chalcopyrite, galena, and rarely enargite may be present.

Umpleby calls attention to the **apparent** playing out of the veins at moderate depth (42). He sug-gests that they only appear to peter out because of erosional effects and their further extension may be expected.

Gold placers occur in the Loon Creek and Yankee Fork districts. In recent years dredging has been done on the Salmon River with considerable production. Umpleby estimates $1,000,000 from placers during the period 1869-1880.

In conclusion he regards northwestern Custer County as a very important area for future prospecting. Lack of transportation at that time was a leading difficulty and still is. Also, vein outcroppings are not prominent, there being a lack of iron stained material. In addition, the region is heavily timbered.

Recent observations suggest this area as deserving a detailed investigation.

ELMORE

Total	Placer		Total		Placer		Total		Placer
1889—14,330			1907— 4,402		200		1925— 5,599	(5,578)	5,545
1890— 3,489			1908— 7,383	(6,795)	238		1926— 3,567	(3,532)	3,422
1891— 4,256			1909— 9,347	(10,166)	166		1927— 3,888	(4,178)	3,859
1892— 5,870			1910— 8,721	(11,528)	149		1928— 4,517	(4,515)	4,504
1893— 2,730			1911—11,528	(11,528)	172		1929— 39	(34)	14
1894— 5,139			1912— 2,957	(3,500)	221		1930— 54	(48)	45
1895— 5,841			1913— 604	(4,920)	224		1931— 279	(244)	154
1896— 3,083			1914— ·969	(2,520)	464		1932—19,164	(19,573)	302
1897— 3,018			1915— 3,730	(3,050)	324		1933—18,017	(17,565)	421
1898— 2,499			1916— 9,627	(11,800)	239		1934—28,642	(31,288)	194
1899— 3,409			1917— 4,755	(4,550)	46		1935—20,009		194
1900— 3,730			1918— 477	(270)	57		1936— 4,349		244
1901— 4,457			1919— 103	(56)	43		1937— 2,722		212
1902— 4,100			1920— 44	(32)	5		1938— 7,103		1,178
1903— 5,103			1921— 72	(55)	22		1939—12,352		549
1904— 3,368			1922— 1,525	(1,525)	1,495		1940—18,685		289
1905— 6,51b		1,126	1923—10,942	(10,942)	10,720		1941—28,505		5,076
1906— 6,394	(7,987)	1,196	1924— 8,822	(8,822)	8,728		1942—29,576		6,410

Total production, 381,396 oz.; placer, 58,447 oz.

Geology.—With the exception of the Snake River placers, the known mineralized area of Elmore County is granite. Fig. 15 (43) gives some information.

Production before 1889 would be credited to Blaine (Alturas) County.

Fig. 2 shows the operation around Rocky Bar and Atlanta in 1882.

According to Anderson most of the ore deposits are along faults and fault or fracture zones that trend from west to slightly north of west. These fractures and fissures are quartz-filled. A very small amount of arsenopyrite, pyrite, sphalerite, galena, chalcopyrite, and sericite is present. Gold is contained in both the quartz and the sulfides. He recognizes three stages of deposition for the quartz and states that gold occurs only in the third stage (44).

The various lodes and veins are made up of one or more combinations of the three quartz stages. The associated minerals, which are peculiar to each stage, makes it possible to distinguish between the three varieties of quartz.

Early or first stage.—This is a massive, milky white, moderately coarse grained variety. It is barren of all other minerals with the exception of rare grains of pyrite. It occurs in many of the veins, in some cases having been brecciated and recemented by later quartz.

Second stage.—The second variety is fine grained and almost chalcedonic, and is distinctly grayish-blue in color instead of milky white. Its color derives from disseminated minute grains of arsenopyrite (which occurs only in second stage quartz) and pyrite. Second stage is not present in all of the veins.

Third stage.—This is the latest quartz. It is distinctly different from the other two. It is white, usually coarsely crystalline, and has a distinct comb and drusy habit. Small amounts of pyrite, sphalerite, chalcopyrite, and galena are usually present. The gold is found in this quartz. Third stage quartz is not present in all of the lodes or veins.

FRANKLIN

No production of gold.

FREMONT

Total	Placer		Total	Placer		Total	Placer
1904— 4			1910—3 (47)	3		1915—3 (57)	
1905— 0			1911—3 (3)	3		1916—0 (125)	
1906—15			1912—3 (3)	3		1917—1 (100)	
1907— 5			1913—5 (27)	5		1918—1 (50)	1
1908— 1 (13)			1914—2 (37)			1920—0	
1909— 7	6						

Total production, 53 oz.; placer, 21 oz.   No gold geology on this county.

GEM

Total	Placer		Total	Placer		Total	Placer
1915—132	12		1925— 19	10		1934— 91	26
1916— 77	6		1926— 4	3		1935— 61	15
1917— 2	1		1927—640 (728)			1936— 183	69
1918— 14			1928—283 (187)	7		1937— 759	567
1919— 54 (170)			1929— 60 (60)			1938—4,049	3,944
1920— 15			1930— 8	8		1939— 151	43
1921— 1			1931— 4	4		1940—1,401	594
1922— 22 (21)	9		1932—456 (325)	21		1941—3,232	611
1923— 19 (19)			1933—144	68		1942—2,917	8

Total production, 14,798 oz.; placer, 6,024 oz.

Geology.—The Boise Quadrangle, Fig. 10, in- cludes a small part of southeastern Gem County. At the time the map was made it was Canyon County.

GOODING

Total	Placer		Total	Placer		Total	Placer
1913— 9 (9)	9		1923— 0 (1)			1936—10	10
1914—31	31		1931— 2	2		1937— 1	1
1915—44	44		1932—45 (27)	45		1938— 8	8
1916—12	12		1933—40	40		1939—19	19
1917— 2	2		1934—41	41		1940— 3	3
1922— 1 (1)			1935— 6	6			

Total production, 274 oz.; placer, 274 oz.   **Geology.**—See Snake River Placers.

IDAHO

Total	Total	Placer		Total	Placer
1862-68—1,500,000	1901— 7,810			1923— 1,155 (924)	402
1874— 9,000	1902— 7,596			1924— 1,801 (1,705)	239
1880—20,738	1903—11,449			1925— 1,598 (1,034)	441
1881—13,000	1904—14,584			1926— 660 (735)	450
1882—12,000	1905—10,543	2,289		1927— 668 (666)	394
1883— 7,500	1906— 8,274 (8,950)	3,371		1928— 1,623 (1,616)	289
1884— 5,500	1907— 5,357	1,964		1929— 2,270 (2,218)	235
1886—19,791	1908— 9,333 (5,744)	2,065		1930— 3,786 (3,564)	153
1887—13,935	1909— 9,927 (5,687)	1,381		1931— 4,534 (4,894)	2,869
1888—15,000	1910— 3,090 (1,236)	889		1932—13,134 (12,344)	8,577
1889—11,500	1911— 3,643 (3,643)	932		1933—25,643 (28,293)	17,432
1890—11,970	1912— 3,975 (2,700)	1,247		1934—24,236 (22,974)	17,418
1891— 9,436	1913— 2,063 (2,640)	909		1935—26,981	18,728
1892— 7,140	1914— 2,115 (2,340)	983		1936—28,960	13,303
1893— 9,140	1915— 4,702 (2,700)	1,361		1937—22,217	9,521
1894—12,116	1916— 6,256 (4,841)	1,034		1938—32,903	17,323
1895—11,791	1917—10,222 (10,588)	523		1939—36,031	20,614
1896— 7,515	1918— 3,970 (3,587)	232		1940—37,061	20,413
1897—11,444	1919— 632	238		1941—37,016	24,357
1898— 9,845	1920— 207 (395)	175		1942—12,774	8,907
1899— 8,048	1921— 970 (1,284)	448			
1900— 7,354	1922— 1,018 (774)	363			

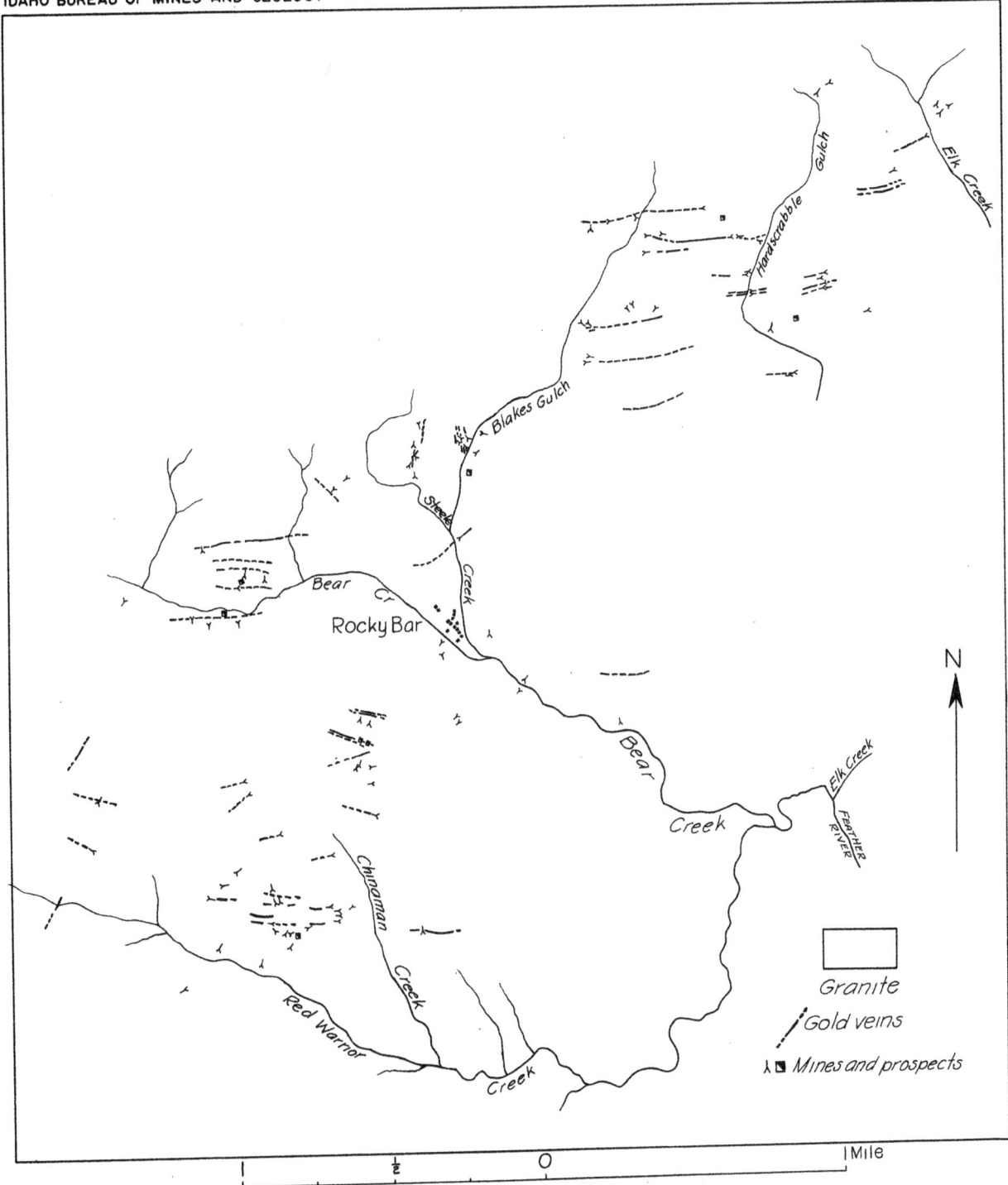

Fig.15—Map of the Rocky Bar district

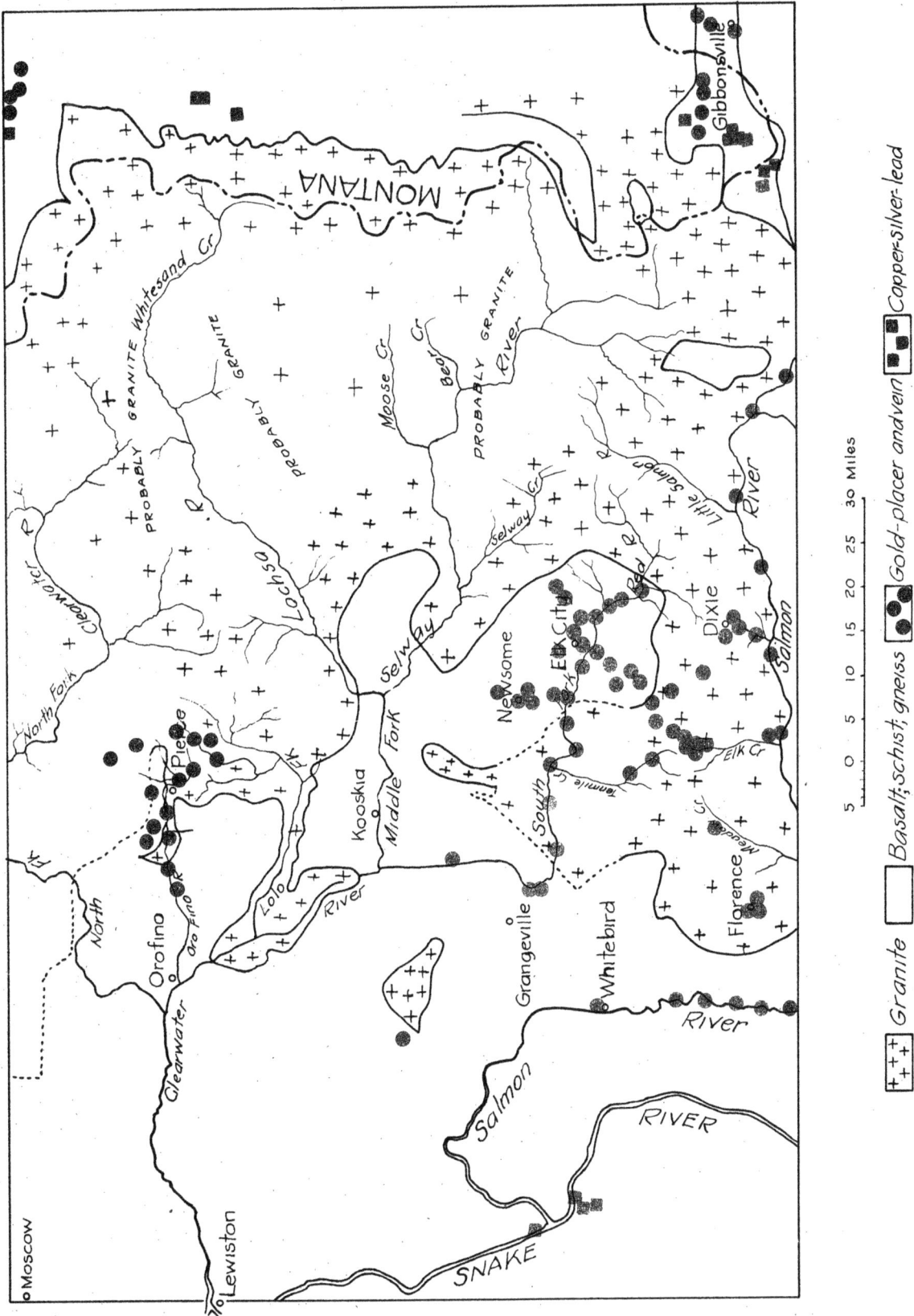

Fig.16–Map of central Idaho from Lewiston to the Bitterroot Range

Total production, 2,176,550 oz.; placer, 202,469 oz.

Geology.—Idaho County ranks second in gold production. It might well lead Boise County if accurate records had been kept. The geology of the area is shown by Fig. 16 (45) and 17 (46). On the Index map it is area three.

An inspection of Fig. 16 shows the wide spread scope of the operations. Just off the map to the south is the Warren area, which was probably responsible for half of the Idaho County production. The workings around Pierce are in Clearwater County.

The gold deposits are almost exclusively fissure veins with extensive placers derived from these veins (47). The veins occur chiefly in gneiss, are quartzose in character and contain from 1 to 10 per cent sulfides in addition to native gold. In some instances the quartz veins occur in granite or slate. Lindgren suggests prospecting in the region between Dixie and Shoup. Shoup is on the Salmon River, just off the map, and is about 25 miles southwest of Gibbonsville.

In the northern part of the county, south of Pierce, the veins are quartz in the schists, and rarely in the granite.

Lindgren (48) summarizes the gold-quartz vein formation as follows:

"Nearly all of the vein deposits occur in granite-gneiss or metamorphic slates and schists. The prevailing strike of the veins seems to be in an east-west direction. The granite, which is the prevailing rock, represents the northward continuation of the great area of central Idaho north of Snake River. Gold bearing veins occur both within this area and along its contacts with the surrounding older sedimentary rocks.

"Within the region here discussed a peculiar relation obtains. The large central areas of granite, whether sheared, as along the eastern margin of the Bitterroot Mountains, or massive, as usually is the case, seem conspicuously barren of deposits. The vein systems appear in or close to the four smaller areas of sedimentary or metamorphic rocks which are found at the periphery of the great central granite area. It is thus in the quartzite series of Lolo Fork, in the quartzites, slates, and gneisses of the upper South Fork of the Bitterroot, and in the old gneiss areas of Elk City and Pierce."

The early production of important areas is as follows:*

Pierce (49)	$ 2,000,000—$ 3,000,000
Elk City (50)	5,000,000— 10,000,000
Florence (51)	15,000,000— 30,000,000
Warren (51)	15,000,000+

* The figures represent the production up to the date of the reference quoted.

See also Fig. 17 for a section of Idaho County.

JEFFERSON

1931—1; 1932—11; 1933—2; 1934—4
Total production, 18 oz.; placer, 18 oz.

Geology.—No information on this county.

JEROME

Total	Placer		Tota	Placer		Total	Placer
1921— 7 (5)	7		1931— 45 (46)	45		1937—149	149
1922— 4 (4)	4		1932— 35 (89)	35		1938—266	266
1923— 0 (1)			1933—162	162		1939—179	179
1924— 8	8		1934— 73	73		1940—289	289
1925— 5	5		1935—111	111		1941—158	158
1926—11	11		1936—196	196		1942— 38	38

Total production, 1736 oz.; placer, 1736 oz.
Geology.—See Snake River Placers.

KOOTENAI

Total	Placer		Total	Placer		Total	Placer
1889—1,500			1908—174	140		1913— 84	84
1904—1,339			1909—100	100		1914—127	127
1905— 3			1910—600	600		1922— 2	2
1906— 97			1911—521 (521)	521		1938— 2	2
1907— 100	400?		1912—300 (300)	300			

Total production, 4,949 oz.; placer, 2,276 oz.

Geology.—From the rather scattered information in the early reports of the Idaho State Mine Inspector (those preceding 1914), the opinion is obtained that the Kootenai production really came from what is now Benewah County. It apparently was almost entirely placer gold from the St. Maries River and probably also the St. Joe River. Since the formation of Benewah County the production has practically ceased.

LATAH

Total	Placer		Total	Placer		Total	Placer
1904—293	225		1917—15 (50)	13		1931— 12	12
1905—225	225		1918— 6 (30)			1932— 34 (8)	24
1906— 37	37		1919—23 (38)	22		1933— 80 (29)	80
1907—357	60		1920—14 (1)	14		1934— 86	86
1908— 30	65		1921—25 (6)	25		1935— 87	87
1909— 40	40		1922—54 (38)	54		1936— 47	47
1910— 58	79		1923—41 (41)	41		1937— 126	120
1911— 46 (46)	46		1924—29 (29)	29		1938— 55	55
1912— 58 (58)	58		1925— 9 (9)	7		1939— 14	14
1913— 22	22		1926—15 (14)	15		1940—2,701	2,694
1914— 34	34		1927—17 (17)	17		1941—5,573	5,573
1915— 32	32		1929— 7	6		1942—6,021	6,021
1916— 36	36		1930— 0	0			

Total production, 16,359 oz.; placer, 15,790 oz.

Geology.—Figure 16 of area 3 takes in southern Latah County. Very little geology, however, is shown. Some placer mining in the vicinity of Moscow has proved attractive in a small way. On Moscow Mountain, in the granite, low grade gold-quartz veins have been investigated. There was at one time a small stamp mill on one of the properties. North and northeast of Harvard are prospects in gold-quartz veins. Considerable early day placering, and very recently dredging, has been successfully undertaken in the Hoodoos a few miles from Harvard.

Pamplet 68   STATE OF IDAHO   April 1946

LEMHI

Total		Total	Placer		Total	Placer
1874— 3,250		1901—10,355			1923— 990 (905)	193
1880— 4,965		1902—10,787			1924— 648 (662)	191
1881—14,250		1903—11,870			1925— 655 (666)	147
1882— 9,000		1904—12,188			1926— 638 (594)	106
1883— 5,000		1905— 4,749	1,440		1927— 697 (682)	142
1884— 3,750		1906— 5,125 (5,179)	1,347		1928— 698 (602)	296
1885— 8,035		1907— 4,074	1,571		1929— 454 (388)	169
1886—16,357		1908— 3,516 (3,415)	1,731		1930— 594 (644)	144
1887—28,299		1909— 6,280 (5,962)	1,164		1931—1,119 (1,041)	171
1888—25,000		1910— 3,239 (1,192)	773		1932—1,323 (1,205)	365
1889—19,000		1911— 5,863 (5,863)	1,622		1933—1,541 (5,091)	694
1890—15,888		1912—10,907 (11,402)	3,923		1934—2,541 (5,091)	694
1891—14,086		1913—13,366 (12,570)	5,162		1935—3,274	626
1892—12,467		1914—14,803 (15,295)	7,656		1936—4,974	698
1893—10,174		1915— 8,817 (7,500)	5,714		1937—4,152	1,036
1894—13,210		1916— 5,470 (5,887)	3,578		1938—6,293	1,132
1895—23,910		1917— 2,884 (3,460)	2,003		1939—6,907	1,177
1896—21,837		1918— 1,975 (1,189)	837		1940—9,487	3,195
1897—15,026		1919— 2,143 (1,812)	165		1941—8,959	4,218
1898—12,569		1920— 1,584 (1,320)	68		1942—6,843	4,531
1899—17,409		1921— 1,383 (1,149)	86			
1900—15,855		1922— 948 (1,085)	171			

Total production, 514,430 oz.; placer, 58,638 oz. A large part of the output in the early years was placer.

Geology.—The geology and location of mines and prospects in Lemhi County are shown in Fig. 18 (53) and 16 (54). On the Index map this is area 5. Additional information may be obtained from other publications (55).

From the map it is seen that Lemhi County contains an exceedingly wide spread mineralization. The small index map of the county is included to show the location of the predominating metals in the county subdivisions.

The geology of the gold deposits of Lemhi County is, according to Umpleby (56), as follows:

The principal ore deposits are fissure veins and replacement deposits along shear zones. In some places the mineralization occurs in joints and crevices in the country rock; in other instances along bedding and joint planes. Again, ore minerals occur as disseminations in the country rock. Two epochs of mineralization are recognized—late Cretaceous or early Eocene and late Miocene or early Pliocene. All the deposits except a small group of gold-silver belong to the Eocene.

Gold-bearing veins are inclosed in many types

S. Fk. of Salmon R.

West Fk

Trout Cr

Sheepeater Pk

Chamberlain Cr

McCalla Cr

Whimstick Cr

SALMON Creek

RIVER

Hungry Cr

Cottonwood Cr

Disappointment

Five Mile Cr

South Fk

Lodgepole

Chicken Pk

S. Fk. of Salmon R

Porphyry Cr

Ramey

Dead Mule Mtn

Cold Mtn

East Fork

Cold Mtn

Beaver Cr

Wolf Fang Cr

IDAHO CO
VALLEY CO

Crooked Cr

Coxey Cr

Cove Cr

Cabin Cr

Black Butte

Wolf Fang Pk

Smith Cr

Creek Cr

Big

Monumental

Horse Mtn

Mt. Eldridge

Logan Cr

Little Marble

EDWARDSBURG

Rush Creek

Profile Pk

Big Cr

Coin Mtn

Cougar Pk

Snowslide Cr

Bear Lake

Lookout Mtn

Rush Cr

Two Point Lookout

Pinnacle Pk

Telephone Cr

West Fk

Profile Cr

Tamarack Cr

Rainbow Pk

Thunder Mtn

YELLOW PINE

East Fk of S. Fk of Salmon River

Rainbow Cr

Coone Cr

Marble Cr

STIBNITE

Anne Cr

Trail Cr

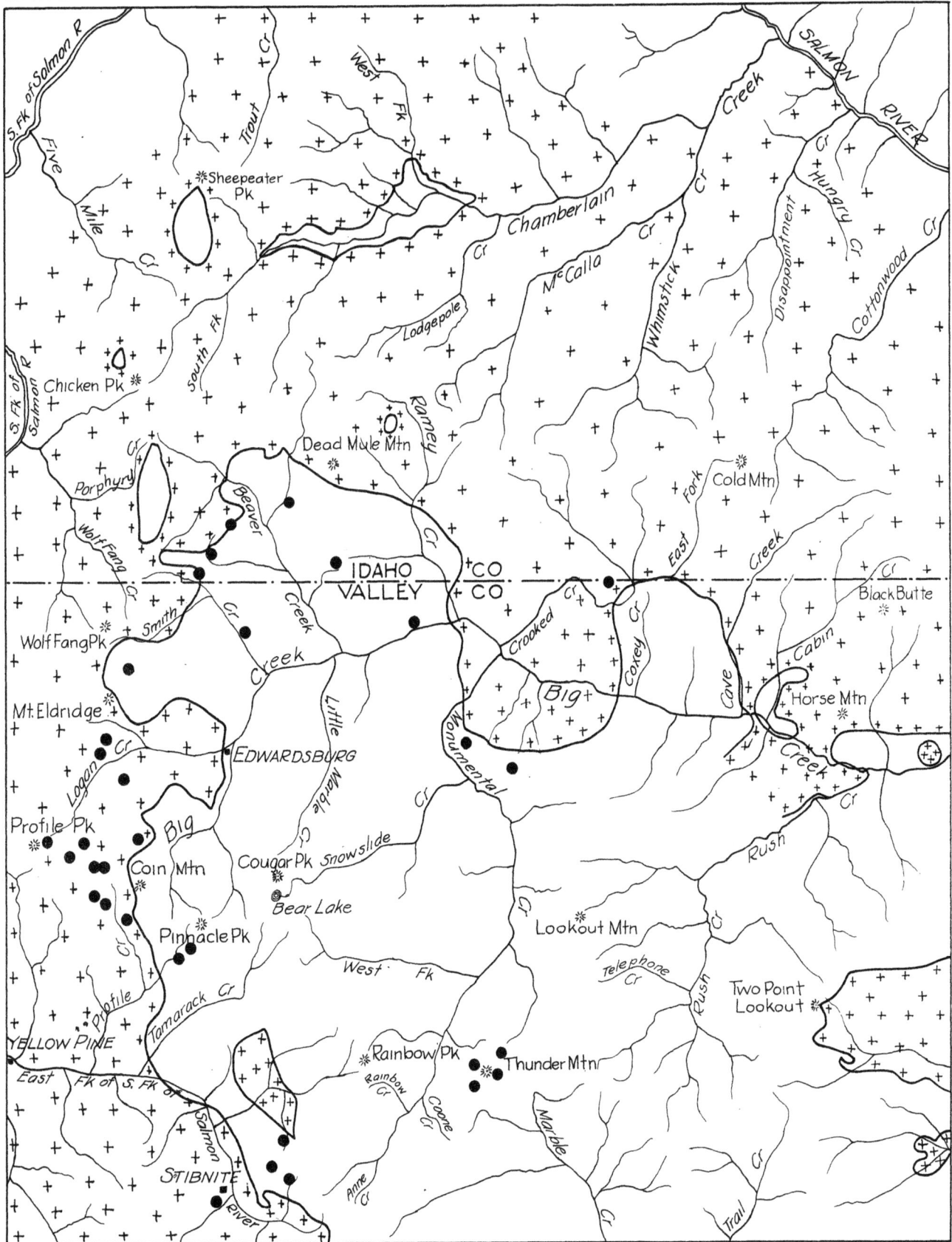

0 1 2 3 4 5 6 7 8 9 10 Miles

+ + + +
+ + + + Granite

Volcanics, sedimentaries ● Mines, prospects

Fig. 17—Geologic map of Edwardsburg and vicinity

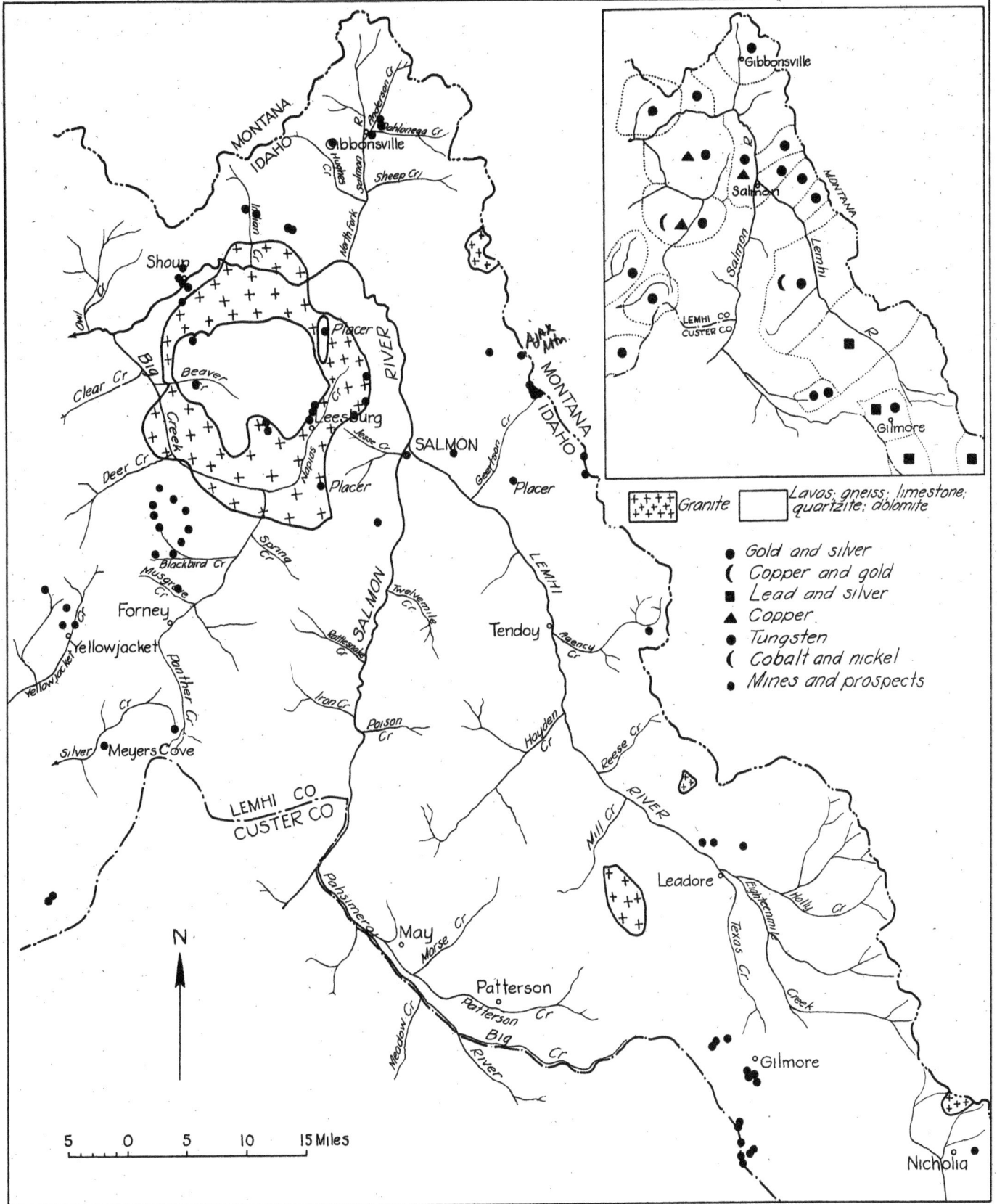

Fig.18—Geologic sketch map of northwestern Lemhi County

of rock in Lemhi County, ore deposits both older and younger than the lavas being clearly recognized. It is noteworthy that the gold is principally in the vicinity of rock of the granite-rhyolite family.

Late Tertiary Gold Veins.—These are associated with eruptive rocks, principally rhyolites. Practically all of them are inclosed in lavas. A few have been found in schists and quartzites, but rhyolite porphyry dikes are also present.

The veins are along fissure fillings and vary in width up to 12 feet and dip at angles greater than 45 deg. The appearance of the Tertiary deposits is distinct from the older deposits. The veins are crustified with wavy bands; there are many unfilled cavities and angular fragments of wall rock encircled by the encrustations.

Adularia, calcite, and sericite with alternating layers of quartz make up the veins. The sulfide minerals are well distributed throughout the vein and are present in minor quantities. Pyrite is the most abundant. Selenium is characteristic of these deposits. These veins are Miocene or later.

Late Cretaceous or early Eocene Gold Veins.— These constitute the most important gold veins of the county. They are the older deposits and are responsible for the major production, both lode and placer.

The veins are inclosed in gneiss, sedimentary rocks, limestone, and granite of varying geologic age. They occur near the contact of sedimentary rocks and granite and may be in either rock. In other cases they are in older rocks traversed by granitic dikes. Some localities show no granite within many miles of the veins. It is thought that the granite is not far below the surface in such cases. This is indicated by the nature of the dikes present.

The strike of the veins varies with the locality. It may be east-west; northeast-southwest; or northwest-southeast. There is nothing constant about the dip. It changes from almost flat to steep. The width averages about 3 feet, but may become 20 feet or more. Faulting in some localities has proved a handicap. The veins pinch and swell; follow parallel joints; and switch from the hanging wall to the footwall of the fissure. The ore solutions followed fractures and cross fractures, selecting the most favorable condition.

These deposits are primarily pyritiferous gold-quartz veins. In some places chalcopyrite replaces the pyrite. Elsewhere both are present.

The vein filling is coarse-textured clear-white quartz. The deposition is bunchy and the bunches irregularly spaced. Pyrite predominates; small amounts of chalcopyrite are usually present; smaller amounts of galena and sphalerite are usually found. Pyrrhotite and arsenopyrite are rare. The mineralization is very irregular.

In Umpleby's (57) opinion there is a close relationship between the granite and the veins. Where the granite appears to be absent, dikes of a nature indicating the presence of the granite at depth are found. The veins may be in either the invaded rock or the granite.

LEWIS

	Total	Placer		Total	Placer		Total	Placer
1932—	17	17	1938—	6	6	1941—	51	51
1933—	7 (6)	7	1939—	31	31	1942—	2	2
1934—	7	4	1940—	107	107			

Total production, 225 oz.; placer, 225 oz.

Geology.—A little general geology is shown by Fig. 16.

The production of Lewis County is apparently all from Salmon River placers.

LINCOLN

	Total			Total	Placer			Total	Placer
1889—	1,500		1898—	1,619			1907—	24	24
1890—	1,048		1899—	1,611			1908—	45 (62)	45
1891—	1,620		1900—	2,108			1909—	43 (27)	44
1892—	776		1901—	1,979			1910—	17 (8)	17
1893—	528		1902—	1,850			1911—	10 (10)	10
1894—	1,407		1903—	459			1912—	38 (3)	39
1895—	1,549		1904—	161			1913—	15	
1896—	843		1905—	156		156	1917—	5	5
1897—	1,498		1906—	137 (121)		137	1921—	0 (1)	

Total production, 21,046 oz.; placer, 477 oz.

Geology.—See Snake River Placers.

Lincoln County was originally Logan County. So far as can be ascertained the production was apparently mainly from Snake River placers.

There has been no production since 1917 which would be about the time that Jerome County crowded Lincoln County from the river. Up to 1911, Lincoln County had a considerable boundary on the Snake River.

MADISON

No production. It seems strange that no Snake River placering had been reported from this county.

MINIDOKA

1914— 0 (350)		1934—95	95
1915— 9	9	1935— 7	7
1933—22	22		

Total production, 133 oz.; placer, 133 oz.

Geology.—See Snake River Placers.

NEZ PERCE

	Total		Placer		Total		Placer		Total	Placer
1880—18,651				1908—2,186	(2,997)	2,186		1931— 5	(33)	5
1881— 250				1909—2,202	(2,457)	2,183		1932—34	(21)	34
1882— 250				1910—1,975	(2,438)	1,980		1933—39	(8)	37
1883— 600				1912— 21	(157)	21		1934—31		24
1884— 25				1913— 170				1935—31		29
1885— 2,650				1914— 1	(270)	1		1936—29		29
1886— 700				1915— 3	(75)	3		1937—28		28
1896— 185				1916— 2		2		1938—72		72
1904— 857				1917— 1		1		1939—94		94
1905— 1,210		951		1918— 6		6		1940—47		47
1906— 2,374	(3,906)	1,861		1921— 0	(49)			1941—39		39
1907— 2,635		2,620		1922— 0	(1)			1942— 3		3

Total production, 37,406 oz.; placer, 12,256 oz.

Geology.—See Fig. 16.

It is practically impossible to say where the early gold credited to Nez Perce County came from. It is one of the original four counties. The boundary was almost continually being changed in the early days. Most authorities say that the Pierce country was in Shoshone County, so the production from there would probably be assigned to that county. There is a persistent declaration by some of the old-timers that the Palouse River near Potlatch, Latah County, produced large quantities of placer gold in the eighties and later. The output for Nez Perce County, in some ways, substantiates this, as Latah County was not formed until the early nineties. Also, the Pierce area and the Oro Fino area probably were credited, at least in part, to Nez Perce County. Since the county boundary became stable, about 1911, the production has been practically nothing; the little gold found has come from placers on the Snake and Salmon Rivers, with perhaps a small part from the Clearwater River.

ONEIDA

	Total		Total	Placer		Total		Placer
1880—4,565		1898—909			1906—194	(197)	194
1881—2,000		1899—750			1907—425		427
1882—1,750		1900—629			1908—213	(202)	215
1883— 400		1901—485			1909— 89	(74)	95
1884— 225		1902—730			1910—101	(52)	101
1895— 732		1903—539			1911— 12	(12)	12
1896— 631		1904—431			1912— 14	(15)	14
1897— 822		1905—393	393				

Total production, 17,089 oz.; placer, 1,451 oz.

Geology.—See Snake River Placers.

The history of Oneida County shows it to have been formed early in the history of the state.

Until Power County was formed about 1912, Oneida contained part of the Snake River. There has been no production since 1912.

For historical data see the reference to Bancroft elsewhere in this report.

OWYHEE

Production previous to 1874 and 1875 to 1879 (58).

1863-1866*, inclusive	$4,000,000 or 150,000 oz. gold	(¾ of output taken as gold)	
1867*	1,000,000 " 25,000 "	" (½ " " " " ")	
1869*	1,600,000 " 40,000 "	" (" " " " " ")	
1870*	842,935 " 21,000 "	" (" " " " " ")	
1871*	981,363 " 25,000 "	" (" " " " " ")	
1872*	455,157 " 12,000 "	" (" " " " " ")	
1873*	1,002,267 " 25,000 "	" (" " " " " ")	
1875*	225,000 " 5,600 "	" (" " " " " ")	
1876-1879*	1,000,000 " 25,000 "	" (" " " " " ")	
Total	328,600 oz.		

* The data for these years did not distinguish between gold and silver. During 1863 and 1864 most of the production was from the Jordan Creek placers. Possibly a major portion of the $4,000,000 shown for 1863-1866 was gold. Lode operations for the early years seems to show 20 to 100 ounces of silver per ounce of gold.

Considerable placering on the Snake River in Owyhee County has been done. Just what the output from this source has been is not known. Bancroft (59) reports it as being substantial. See also the Snake River Placers.

Total	Total	Placer	Total	Placer
1874—45,000	1901—38,912		1923— 314 (296)	0
1880— 2,961	1902—20,938		1924— 291 (191)	2
1881— 2,500	1903—20,479		1925— 413 (426)	0
1882—10,000	1904—23,584		1926— 190 (191)	0
1883— 7,500	1905—14,866	68	1927— 239 (209)	6
1884— 5,000	1906— 9,493 (8,645)	104	1928— 861 (914)	778
1885— 6,747	1907—17,218	66	1929—1,217 (1,206)	947
1886— 3,318	1908—18,862 (18,595)	54	1930— 909 (911)	660
1887— 4,541	1909—20,574 (19,593)	83	1931— 703 (711)	546
1888— 5,765	1910—18,805 (18,696)	31	1932— 634 (638)	468
1889—12,260	1911—15,587 (15,587)	66	1933— 708 (66)	397
1890—16,556	1912—14,031 (14,270)	38	1934—1,653 (983)	325
1891—16,254	1913— 8,139 (7,125)	71	1935—2,139	1,942
1892—23,244	1914— 861 (1,000)	75	1936—3,979	3,643
1893—25,344	1915— 2,477 (3,017)	12	1937—4,807	4,605
1894—37,915	1916— 500 (550)	9	1938—7,726	6,856
1895—35,204	1917— 617 (600)	1	1939—6,548	2,997
1896—32,948	1918— 423 (900)	3	1940—7,694	2,728
1897—33,666	1919— 208 (424)	6	1941—5,850	751
1898—34,275	1920— 108 (600)	0	1942—3,548	18
1899—34,001	1921— 209 (215)	7		
1900—37,538	1922— 243 (253)	5		

Total production, 730,094 oz.; placer, 28,368 oz. The total, with the estimated output for the early years, is 1,058,694 oz.

Geology.—For geological maps see Fig. 19 (60), 20 (60), and 21 (60).

Dr. F. A. Thomson, then Secretary of the Idaho Bureau of Mines and Geology, in the preface to the Piper and Laney report, expresses the opinion that careful prospecting will show extension of the ore, laterally and at depth, beyond the operations known at that time (1926). Practically no new work has been done since then. A study of Fig. 19, 20, and 21 seems to bear out his assertion.

The structure of the Silver City region involves a complex system of block faulting. This was overlooked in early reports on the district (61). Before planning extensive development the report by Piper and Laney should be consulted. There appears to be a definite connection between the various periods of fracturing and ore deposition.

Three major periods of fracturing have been observed. They are those connected with the granite (actually a granodiorite) intrusive; the one following the extrusion of the rhyolites; and the fractures following ore deposition.

The fracturing or faulting related to the granite did not affect the overlying basalts and rhyolites.

Fracturing following the rhyolite is somewhat complex. The veins have been offset and rotated from their original strike. Numerous sets of secondary fracturing have caused rotation of the veins. The fractures which strike N. 75° W. and N. 25°-40° W. are responsible for the greatest displacement. As much as 2,000 feet has been observed.

The last set of disturbances are post-mineral fractures. The "iron dike" in the DeLamar district is the outstanding representative. It is important to recognize that the so-called "iron dike" is a post-mineral fault and not pre-mineral as formerly thought (62). The veins have been fractured and displaced by this fault.

In summarizing the fracturing it is well to quote directly from the authors (63):

"Inasmuch as the fault systems which have been described will constitute the critical problem to be met in any future search for ore bodies, it is essential that the problem be clearly recognized. With the exception of the DeLamar and other parallel faults, the displacements to be expected on War Eagle, Florida, and DeLamar Mountains are small and their directions may usually be predicted. Not all the fractures that exist have been detected, of course, but others can probably be classified in one of the sets already described. Along the faults of the "iron dike" system the displacements are probably as great as 1,200 feet or more, and careful search is required before the extension of the vein systems will be found. The problem is quite different in the Flint district inasmuch as the ore deposits are genetically related to the first period of diastrophism (fracturing) and have been deformed in each of the two succeeding periods. The displacements are, therefore, relatively large. Moreover, several parallel veins may be displaced so that the two segments most nearly in alignment in the opposing walls of a fault may not be portions of the same vein or ore body. The analysis of the fault problem becomes, therefore, of critical importance. Competent geologic guidance accompanied by detailed geologic mapping, which attains its greatest value only when it keeps ahead of the timberman, becomes an essential part of the exploratory program of any magnitude."

The veins.—All of the veins in the region are filled fissures (64). There are four types (64):

(1) Fissure filled with massive ore-bearing white or milky quartz with varying amounts of crystallized quartz. (This is typical of the Flint district.)

(2) Fissure filled with lamellar, pseudomorphic quartz. (This is typical of the DeLamar district.)

(3) Silicified and mineralized shear-zones, of which the Poorman vein appears to be a typical example (on War Eagle Mountain, Silver City district.)

(4) Cemented breccias, the cementing material being either massive and crystallized quartz, or pseudomorphic quartz; the Oro Fino-Golden Chariot vein, and portions of the Seventy-seven vein at DeLamar are examples of this class.

Some of the veins show more than one of the characteristics listed above.

Essentially the gangue and ore minerals of all the districts are the same. The difference is mainly in the abundance of certain minerals.

Many of the veins are in very strong fissures and have been developed as much as 6,000 feet on the strike and 1,800 feet in depth.

In investigating this district it should be kept in mind that the ore occurs in well defined shoots (65). That is, more or less alternating portions of the vein will be productive and barren. All intersections of fractures and shear zones with the vein should be investigated. The shoot may extend vertically or horizontally or both. Mining should not be stopped simply because a given ore shoot becomes exhausted.

In conclusion it may be pointed out that around Silver City (Fig. 19) the veins may outcrop in granite, rhyolite and basalt; at DeLamar (Fig. 21) they occur in rhyolite but might well extend downward into the underlying granite; and at Flint (Fig. 20) in the granite, but the surrounding rhyolite and basalt may contain vein outcrops.

PAYETTE

	Total	Placer
1933—	7	7
1934—	3	3

Total production, 10 oz.; placer, 10 oz.
Production is apparently from the Snake River.

POWER

	Total	Placer		Total		Placer		Total	Placer
1913—	58	58	1924—	2		2	1936—	51	51
1914—	34 (60)	34	1925—	8	(39)	8	1937—	698	698
1915—	34 (9)	34	1926—	8		8	1938—	24	24
1916—	78	78	1931—	4		4	1939—	32	32
1917—	11	11	1932—	66	(46)	66	1940—	19	19
1918—	10	10	1933—	36		36	1941—	18	18
1921—	6 (11)	6	1934—	168		168	1942—	0	0
1922—	8 (4)	8	1935—	73					

Total 1,446 oz.; placer, 1,446 oz.

Geology.—Power County was taken from Oneida County. The production is from Snake River plac-

ers and was first recorded for 1913. The year 1912 showed the last record for Oneida.

See Snake River Placers.

Granite

Rhyolite and Basalt

Gold-silver veins

War Eagle Mtn

SILVER CITY

Jordan

Creek

Creek

Jordan

Long Gulch

Florida Mtn

Sawpit Creek

N

1 Mile

0

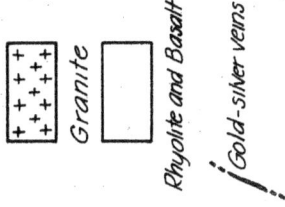

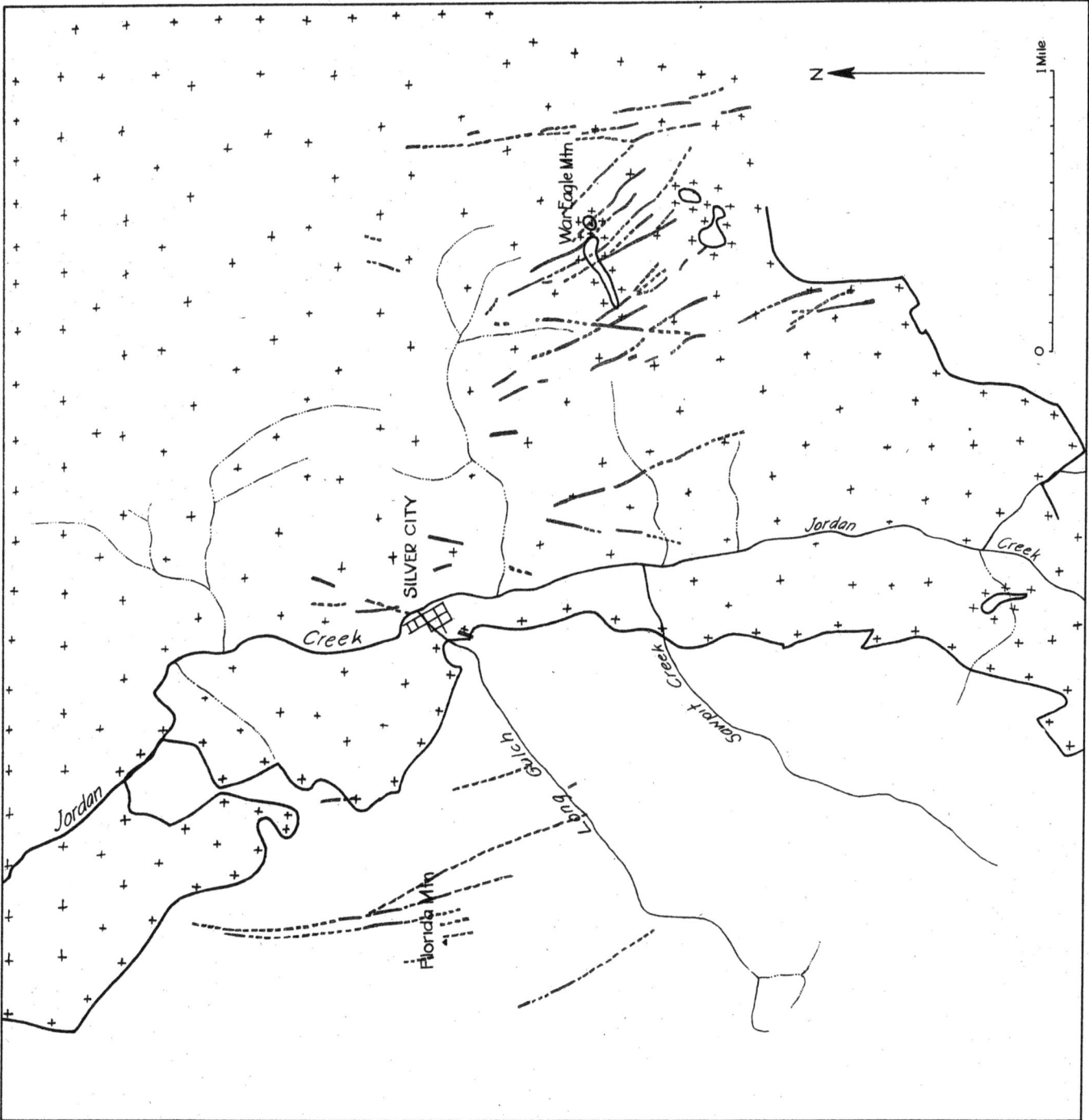

Fig.19-Geologic map of the Silver City area

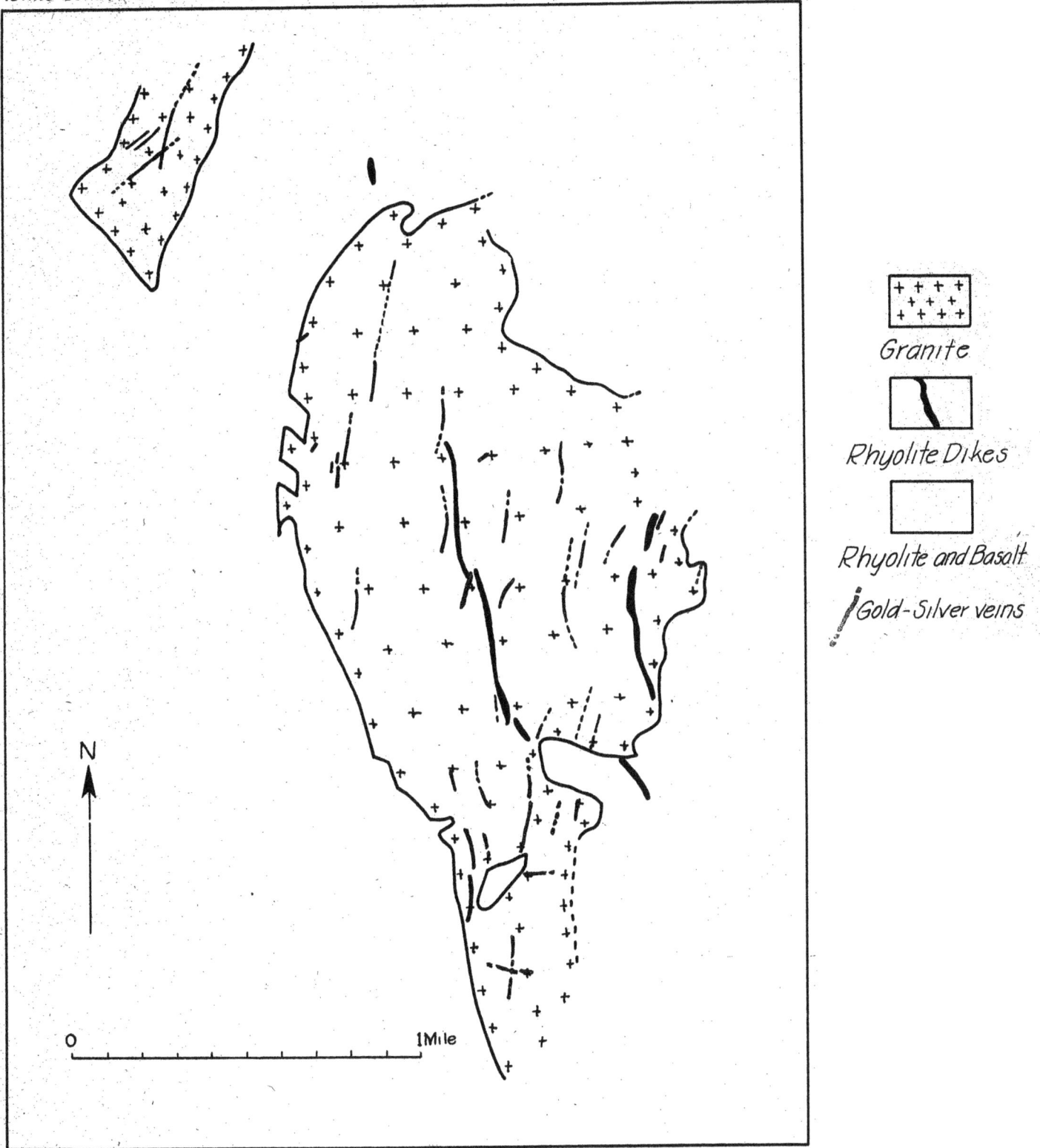

Granite

Rhyolite Dikes

Rhyolite and Basalt

Gold-Silver veins

N

0   1 Mile

Fig. 20-Geologic map of the Flint mining district

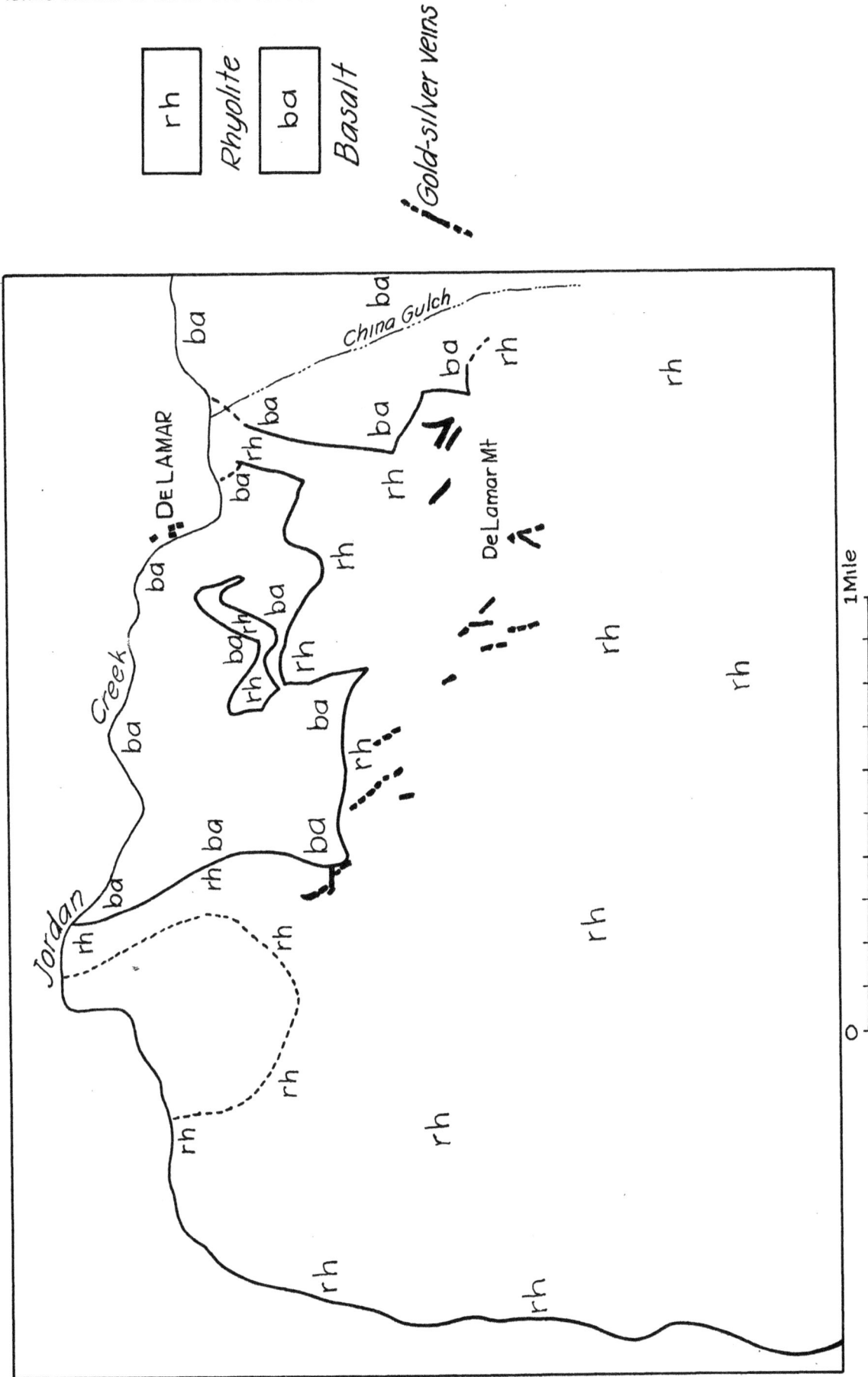

Rhyolite rh

Basalt ba

Gold-silver veins

DELAMAR

China Gulch

DeLamar Mt

Jordan Creek

1 Mile

0

Fig.21-Geologic map of the DeLamar mining district

Prichard Formation
(Argillite, slate, and gray
and white quartzite)

Remainder of Belt
Series gravels, and
placer tailings

● Mines, prospects

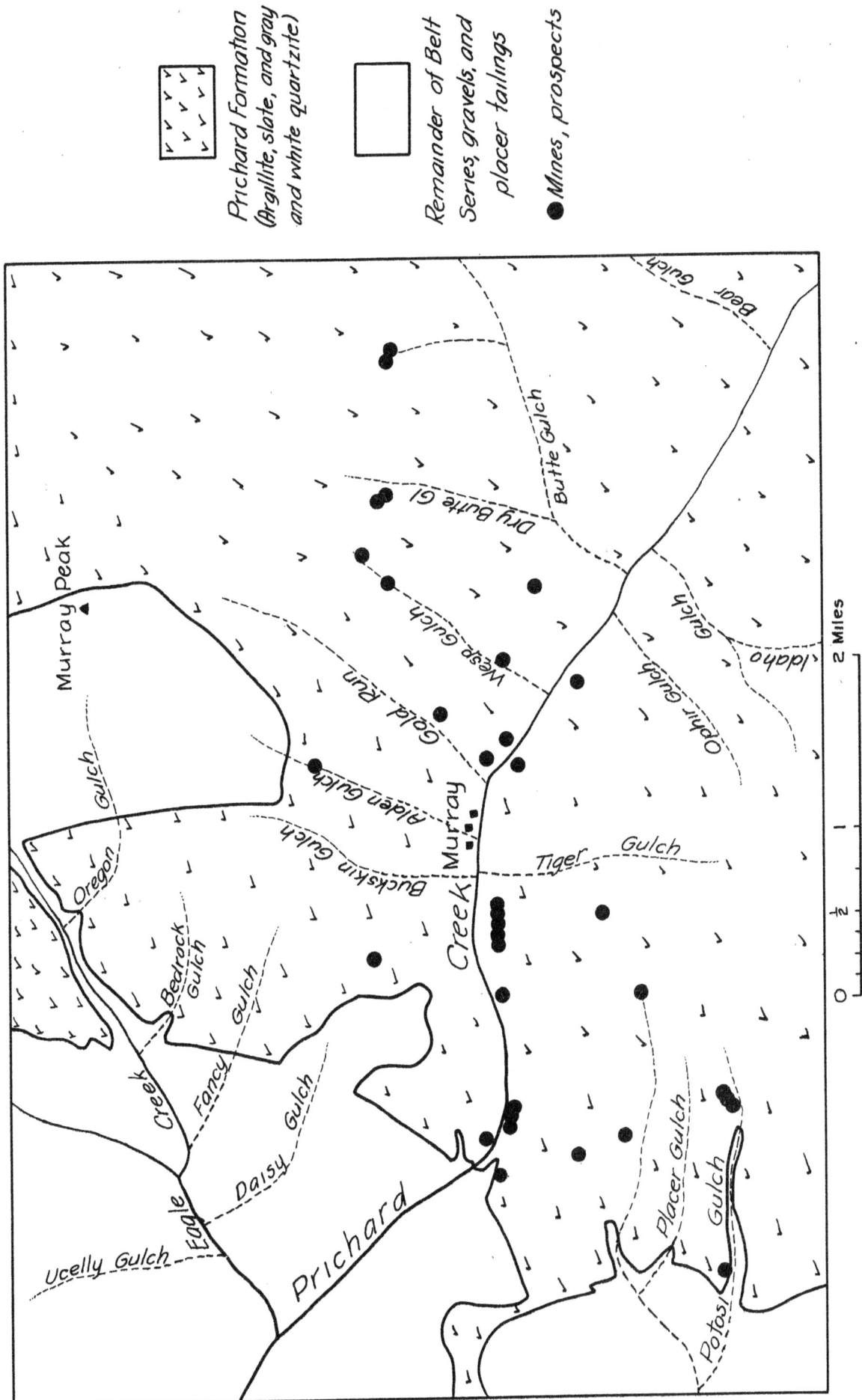

Fig.22-Geologic map of the Murray district

SHOSHONE

Total	Total	Placer	Total	Placer
1881— 3,000	1902— 4,761		1923—13,182 (11,873)	11,784
1882— 2,500	1903— 7,651		1924— 8,602 (8,026)	7,812
1883— 2,500	1904— 2,226		1925— 6,615 (6,260)	5,766
1884—12,500	1905— 1,886	1,640	1926— 3,441 (5,060)	2,243
1885—15,048	1906— 4,190 (3,244)	621	1927— 416 (436)	52
1886—11,983	1907— 3,952	834	1928— 428 (441)	64
1887— 7,367	1908— 3,878 (4,105)	585	1929— 511 (637)	38
1888—10,250	1909— 4,326 (7,156)	667	1930— 564 (811)	69
1889— 8,433	1910— 3,148 (3,110)	436	1931— 456 (673)	61
1890— 8,000	1911— 4,162 (4,162)	515	1932— 395 (399)	175
1891—10,000	1912— 4,085 (4,447)	365	1933— 1,585 (1,001)	587
1892—11,000	1913— 3,950 (3,510)	242	1934— 3,965 (1,477)	851
1893—14,748	1914— 3,104 (4,052)	100	1935— 2,714	734
1894—17,531	1915— 2,246 (4,500)	90	1936— 2,454	489
1895—18,439	1916— 2,247 (2,600)	105	1937— 3,659	732
1896—17,369	1917— 4,145 (4,753)	576	1938— 4,053	284
1897—16,402	1918—11,873 (14,628)	9,015	1939— 5,928	578
1898—13,011	1919— 8,687 (10,191)	7,568	1940— 6,886	1,933
1899— 8,602	1920— 5,897 (9,000)	4,633	1941— 3,419	363
1900— 5,754	1921— 8,306 (7,217)	7,202	1942— 2,688	62
1901— 4,915	1922— 7,056 (6,049)	5,950		

Total production, 393,088 oz.; placer, 75,821 oz.

Geology.—Shoshone County was one of the original founder counties. The location of the boundary with Nez Perce on the south depends on whose map is consulted. Plate 1 gives some indication of its indefiniteness. Previous to the beginning of this century, the boundary was apparently in a continuous state of change. Before 1881, therefore, gold production assigned to Shoshone County probably came from Nez Perce, and later Clearwater County, as the production from Murray did not start until about 1882. That is, the Pierce area, and possibly minor quantities from the St. Joe River area would be credited to Shoshone County. After 1882 the Murray district provided most of the straight gold output. The bulk of the gold came from the copper and lead ores (66).

On the Index map are shown two definite gold producing areas for which maps are available. Fig. 22 (67) and 23 (68) reproduce these areas.

In addition to the areas shown, a little gold has been found on Elk Creek, a few miles east of Kellogg, and in Gold Run Gulch, 2½ miles east of Kellogg (69).

Murray district (Fig. 22).—The gold-quartz veins in the Murray district occur in the Prichard slate. There are three types of these veins:

(1) Mineralized shear zones that cut across the bedding of the inclosing rock, usually at steep angles.

(2) Quartz veins that lie approximately along the bedding of the inclosing rocks.

(3) Quartz veins that lie along low-angle thrust faults.

The first class are widely distributed. They lie along well defined and persistent faults. The ore and gangue minerals are unequally distributed along the shear zones. The principal gangue minerals are quartz, altered wall-rock, ankerite, pyrrhotite, pyrite, galena, sphalerite, chalcopyrite, and in some places, arsenopyrite. Some free gold is present.

The bedding veins occur throughout the Murray district. Many of them are too small or too low-grade to work. They lie along the bedding of thinly laminated dark gray argillite (clayey slate); although the veins occasionally leave the bedding and follow fractures. There are usually two or more parallel bedding veins. The minerals present are, arsenopyrite, pyrite, galena, chalcopyrite, specularite, scheelite, and gold. Selenium is reported.

The gangue is mostly white quartz with lesser amounts of ankerite, sericite, etc. Attempts have been made to save the scheelite.

There is only one example mentioned of the thrust fault type of vein (71). The mineralogy is very similar to that in the bedding veins.

There are numerous bench and stream placer deposits. Many of them have been worked. Others, because of the lack of water, and consolidation of the gravel have proved unprofitable in the past. Shenon (72) believes there are many of them worth investigating. In addition he takes an optimistic view of future lode mining. The past production of $2,000,000 (up to 1937) has come from those veins readily found. The area is densely forested and future prospecting stands a good chance of exposing more productive veins.

St. Joe River area (Fig. 23).—Prospectors entered the area about 1873 (73) and discovered gold.

This territory has been very sketchily geologized. It needs more detailed work. Fig. 23 shows two granite areas. The one around Black Prince Creek has favorable possibilities according to Pardee.

The lodes that have been investigated are made up of quartz, siderite, and calcite, with chalcopyrite as the main ore mineral. Some of the veins contain galena and sphalerite. More or less pyrite is present in most cases.

Very little is known about the area but it would appear to be worth prospecting.

TETON

No production from this county.

TWIN FALLS

Total		Placer	Total		Placer	Total		Placer
1907—110		110	1919— 3	(119)	3	1931— 30	(35)	30
1908—107	(108)	108	1920— 3		3	1932—201	(67)	201
1909—130	(149)	130	1921— 8	(1)	8	1933—114		114
1910— 75	(73)	75	1922—34	(34)	34	1934—169		169
1911— 99	(99)	99	1923— 8	(6)	8	1935—125		125
1912— 22	(127)	22	1924— 9		9	1936—104		104
1913— 45	(22)	45	1925— 8		8	1937— 48		48
1914— 27	(70)	27	1926—54		54	1938— 85		85
1915— 80	(17)	80	1927— 8		8	1939—132		132
1916— 25	(87)	25	1928— 2		2	1940—188		188
1917— 12		12	1929—20	(14)	20	1941—202		202
1918— 15		15	1930— 8		8	1942— 37		37

Total production, 2,347 oz.; placer, 2,348 oz.

Geology.—See Snake River Placers.

VALLEY

Total		Placer	Total		Placer	Total	Placer
1917— 34		3	1926— 384	(212)	109	1935— 9,135	90
1918—236		221	1927— 188	(160)	96	1936— 8,336	190
1919—100		100	1928— 183	(161)	76	1937— 6,379	167
1920— 72		72	1929— 229	(232)	17	1938— 4,832	116
1921— 77	(52)	55	1930— 955	(990)	36	1939— 6,297	167
1922—981	(651)	18	1931— 733	(734)	11	1940—13,225	258
1923—768	(752)	43	1932— 7,322	(7,322)	31	1941—10,346	57
1924—114	(109)	9	1933—10,810	(10,527)	49	1942— 3,708	48
1925— 40	(21)	24	1934—11,094	(11,555)	113		

Total production, 96,578 oz.; placer, 2,176 oz.

Geology.—Part of area 4, Fig. 17 (74) gives information on Valley County. The northern part of the map includes Idaho County. The remarks to follow will cover the entire figure.

Valley County is one of the more recently established subdivisions of the state and was formed from southern Idaho and northern Boise Counties. The production dates from 1917 but no doubt a respectable amount of the Idaho and Boise County totals came from this area.

There is considerable placer ground that has been left untouched (75) because of boulders, almost total lack of transportation, and severe climatic conditions.

There are four kinds of gold lodes:

(1) Lodes containing arsenic.
(2) Lodes containing antimony.
(3) Lodes in pre-Tertiary rocks.
(4) Lodes in Tertiary rocks.

The first are low-grade gold ores carrying auriferous arsenopyrite and pyrite (76). These lodes are wide, silicified shear-zones in granite country rock. Small amounts of pyrrhotite, chalcopyrite, and molybdenite have been observed. Silicification of the highly shattered granite country rock is characteristic. In some of the lodes there is abundant stibnite.

The gold-antimony veins contain stibnite as the chief ore mineral. They occur in the quartz monzonite phase of the granite and in the quartzite country rock (77). Occasionally this type occurs with the first type. Usually, though, the gold-antimony veins appear in narrow shear and fracture zones, and contain more metallic minerals than the gold-arsenopyrite deposits. Pyrite is present. The gangue is quartz and carbonate minerals. Occasionally some cinnabar is observed.

Pre-Tertiary.—The gold deposits of the pre-Tertiary rocks are essentially of two classes; large mineralized shear zones with small amounts of gold; and quartz veins with shoots of relatively

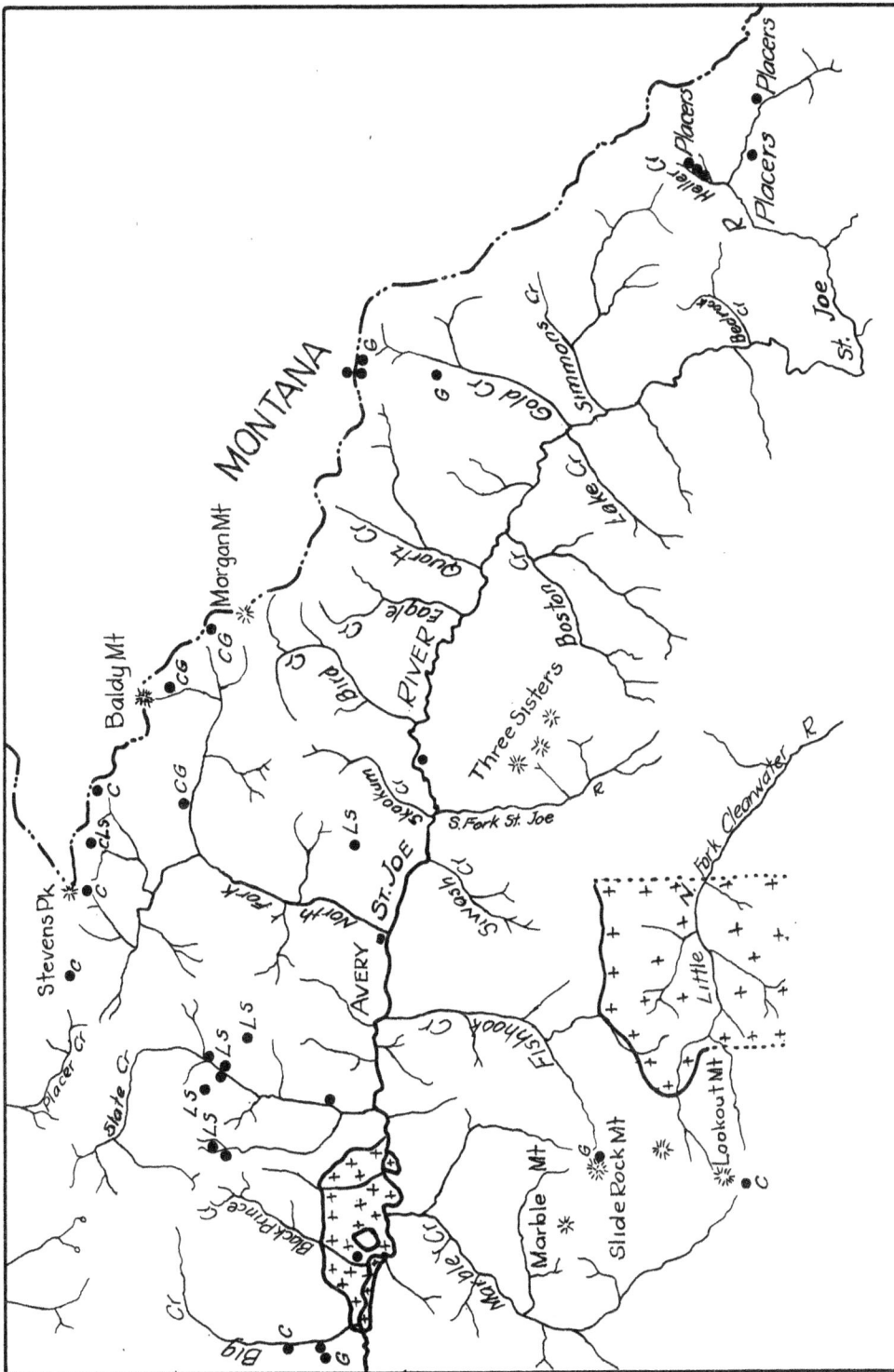

Fig.23-Sketch map of St. Joe River district

good grade ore. The shear zone deposits occur in both the granitic rocks and in the invaded Belt series (metamorphosed sedimentary rocks—quartzite, slate, limestone, gneiss). At the vein contact the rock is silicified, bleached, and sericitized. These zones are made up of a large number of small quartz veins. Associated with the gold in more or less small amounts are pyrite, sphalerite, galena, chalcopyrite, fluorite, tetrahedrite, arsenopyrite, stibnite, and molybdenite. The sulfides at the surface show little oxidation. The gold content is highest where the small quartz veins contain fine-grained bluish-white quartz.

Tertiary Rocks.—These deposits are in the Challis volcanics in the vicinity of Thunder Mountain. The Challis volcanics are described by Shenon and Ross as follows:

"The greater part of the formation is composed of flows ranging from andesite to quartz latite. Beds of white tuff and rhyolitic rocks, which constitute the larger part of the upper members of the formation, are found in this region mainly in and immediately around the Thunder Mountain district . . . . There is a thickness of over 1,000 feet of light-colored rhyolitic lava, flow breccia, tuff, and tuffaceous conglomerate, all containing in places fragments of charred and silicified wood and all somewhat kaolinized . . ."

The ore deposits are of considerable lateral extent but do not gain very great depth. It is very difficult to tell the ore from the altered country rock; sampling and assaying is the only sure way. There are very few sulfides, pyrite being the most abundant. The ore is relatively low grade, although in some places there has been some secondary enrichment. The ore may occur in altered rhyolitic tuff, sandstone, rhyolitic lava, flow breccia, and related rocks. Gold deposits in the Challis volcanics are hard to distinguish from the altered country rock in which they occur. It is very easy to overlook valuable deposits as compared to the pre-Tertiary rocks, where the gold and other deposits are in quartz veins or shear zones replaced by quartz. Crushing and panning and assaying must be relied upon. The favorable areas are those of intense alteration with the formation of abundant clay minerals.

Many of the deposits in the Challis volcanics have been displaced by land slides. This should be kept in mind when trying to follow a deposit along its lateral extent.

WASHINGTON

	Total			Total	Placer			Total	Placer
1886—	1,342		1905—	1,484	35		1924—	12	12
1888—	1,000		1906—	429 (350)	263		1925—	151 (60)	26
1889—	500		1907—	63	2		1926—	31 (20)	29
1890—	604		1908—	4 (150)			1927—	2	2
1891—	810		1909—	11 (165)			1928—	27 (13)	27
1892—	875		1910—	174 (56)	25		1929—	2	
1893—	728		1911—	8 (8)	8		1930—	80 (75)	
1894—	750		1912—	66			1931—	3	
1895—	435		1913—	350			1933—	21 (5)	20
1896—	329		1914—	540			1934—	17	12
1897—	379		1915—	19			1935—	94	77
1898—	434		1917—	0			1936—	15	
1899—	631		1918—	0 (17)			1937—	3	
1900—	551		1919—	0 (37)			1938—	6	5
1901—	286		1920—	0 (20)			1939—	39	33
1902—	341		1921—	4 (42)			1940—	48	36
1903—	1,312		1922—	8	6		1941—	15	6
1904—	2,173		1923—	23	16		1942—	3	1

Total production, 17,232 oz.; placer, 641 oz.

Geology.—Washington County has not been noted for straight gold production. The bulk of the gold has probably been derived as a by-product of the limited copper mining and lead-silver mining. There are some large quartz veins said to carry gold near Bear (now in Adams County). During the peak years, the present Adams County was northern Washington County. Since 1911 the production of Washington County has dropped very materially. There are several granitic areas that should be prospected.

The placer gold is probably, at least recently, almost entirely from the Snake River placers.

SNAKE RIVER PLACERS

Figure 24 (79) *, shows the portion of the Snake River in Idaho that has proved productive. In more recent years, and no doubt the earlier ones also, mining has been done as far north as Lewiston.

Production.—The estimates of the total gold available from the Snake River sands and gravels reaches astronomical figures; it has been placed as high as several billion dollars. The actual total production is almost as astonishing. The estimates by former investigators are much too low. They apparently overlooked the rapid formation of counties in south Idaho and, therefore, failed to credit the river with its just due. Counties like Owyhee, Alturas, and Logan (see Plate 1) had gold production from veins, regular gravel placers, and the Snake River. The Snake River production was lost in the shuffle. It is now impossible to say what proportion of the total should be credited to the river. Also, the placer records are not available previous to 1905. As there is no known, or at least important, vein mining in most of the counties bordering the river, the gold production from such counties must have come from the Snake River sands. In addition a certain amount from Owyhee County and Lincoln County (which no longer has

a border on the river) must be credited to the total. The estimate to follow is, therefore, at best approximate. But on the other hand, it gives a good idea of the importance of this source.

There would appear to be areas along the river which could be mined by draglines, the sand and gravel passed over trommels, and the gold recovered from the undersize sands by flotation. Dredging has in one instance been successful (84). Careful geological investigation, sampling for yardage, and cost calculations would have to be made.

No doubt after each flood season there is an apparent concentration of fine gold along the river. The agitation of the sands by the rapidly moving water and their redeposition may cause the exposure of pay-sand or the reconcentration in favorable places (wide spots, bends, and bars).

The counties at present, which either border on or contain portions of the Snake River are: Nez Perce, Idaho, Adams, Washington, Payette, Canyon, Ada, Elmore, Owyhee, Gooding, Twin Falls, Jerome, Minidoka, Cassia, Power, Bingham, Bannock, (just barely touches), Bonneville, Jefferson, and Madison. Of these, the first five are usually not considered. Hill includes only those above the junction with the Boise River. In this report all will be included. And in addition, several which no longer touch the river and whose gold production ceased at the time the river boundary disappeared.

The following table gives the total production. For individual years consult the county tables.

---

* This is a very excellent presentation of this subject. The location of areas, geology, origin, mining methods and equipment, and mineral analysis of the sands is discussed. See also Pamphlet 72, Ida. Bur. of Min. and Geol., Fine Gold of Snake River and Lower Salmon River, Idaho.

	Production	Remarks
Ada	2,139*	Impossible to say. 1% of recorded data taken.
Adams	830	Impossible to say. 1% of recorded data taken.
Bannock	4,200	
Bingham	24,240	
Bonneville	2,864	
Canyon	623*	Impossible to say. 10% of total taken.
Cassia	22,000	A little vein gold in recent years. By far the greater part must be Snake River gold.
Gooding	273	
Elmore	759*	Impossible to say. 0.2% of total taken.
Idaho	500*	Impossible to say. Very insignificant compared to total.
Jefferson	18	
Jerome	1,736	
Lincoln	14,124	From 1895 to date, probably all Snake River. Before then impossible to say.
Madison	0	
Minidoka	133	
Nez Perce	17	Impossible to say. Half of output since 1911 taken. Probably much higher than shown.
Oneida	17,039	Some doubt as to whether this is all Snake River (80) †.
Owyhee	3,521	Impossible to say. Equivalent of Ada, Canyon, and Elmore taken.
Payette	10	
Power	1,446	
Twin Falls	2,347	
Washington	813*	Impossible to say. 5% of total taken.
Total, ounces	99,632	

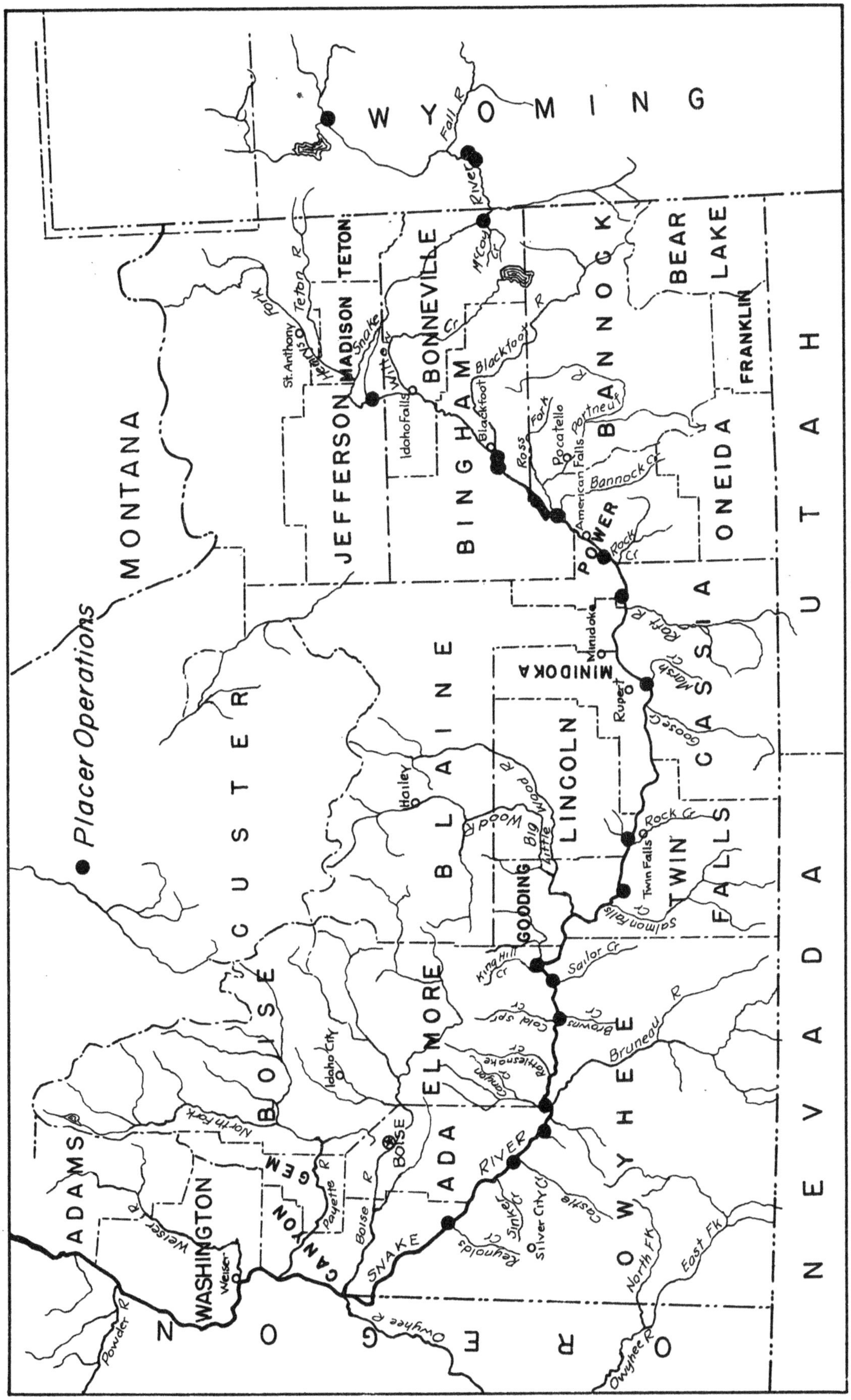

Fig.24-Sketch map of southern Idaho showing main tributaries of Snake River and principal localities at which placer work has been done

There is no substantiating evidence for the assumptions made in the above table. If those marked * are left out entirely, the total becomes 91,277 oz. When it is realized that mining on the Idaho part of the Snake River started in 1872 (81), and that figures before 1880 are not known, the assumptions probably do not give a picture too far wrong. In fact, the total shown may be substantially low.

Bancroft (82) says—

"On the gravel in the vicinity of the Great Falls, at the mouth of Raft River, Henry's Ferry (between the mouths of Bruneau River and Castle Creek), and the mouth of Catherine Creek . . . In 1871 and 1872 several mining camps or towns sprang up along the river and thousands of ounces of gold dust of the very finest quality were taken from the gravels in their neighborhood in these two years. The placers, however, were quickly exhausted on the lower bars, the implements in use failing to save any but the coarsest particles. The highest bars were unprospected, and the camps abandoned. But about 1879 there was a revival of interest in the Snake River placers and an improvement in appliances for mining them and saving gold, which enabled operators to work the high bars which for hundreds of miles are gold bearing."

The placer deposits are of both stream and bench type. The stream placers consist of boulders, gravels, and sands, and form bars, banks, fills, and shoals along the present stream (83). They change their position during flood conditions. Bench placers are older stream deposits, at higher levels represented by terrace remnants (83). Many of the bars in the river have been the best producers. A little platinum is occasionally found.

Hill's article should be studied as a guide toward extensive prospecting.

_____

† Bancroft mentions a production of $250,000 per year for several years as coming from the Cariboo gold district (more recently known as Mt. Pisgah). Some of this is from placers on the eastern slope of the mountain around McCoy Creek. Copper ore was also shipped from here. He does not say the above value is all gold.

REFERENCES

1. Hailey, John, The History of Idaho, p. 29 (1910).

2, 3. Idem, p. 36 and p. 41.

4. Idem, p. 45.

5. Hill, J. M., The fine gold of the Snake River, Idaho. U. S. Geol. Sur. Bull. 620, p. 271 (1916).

6. Hailey, John, Idem, p. 65.

7. Idem, p. 149.

8. Idem, p. 78.

9. Executive Document, 2nd Ses., 47th Cong., vol. 24, Nos. 106 and 108, 1882-83, p. 195, section on Idaho.

10. Bancroft, H. H., Works of, History of Washington, Idaho, and Montana, vol. 31, fn. 62, p. 440 (1890).

11. Idem, p. 535.

12. Emmons, W. H., Gold deposits of the world (1937) (McGraw-Hill Book Co., New York).

13. Bateman, A. M., Economic mineral deposits (1942) (John Wiley & Sons, New York).

14. Newhouse, W. H., (Ed.), Ore deposits as related to structural features (1942) (Princeton University Press, Princeton, N. J.)

15. Newhouse, W. H., Idem, p. 41.

16. Idem, p. 45.

17. Idem, p. 35.

18. Bateman, A. M., Idem, p. 421.

19. Emmons, W. H., Idem, p. 12.

20. Ore deposits of the Western United States, American Inst. of Min. and Met. Engrs., pp. 329 and 330 (1933) (New York).

21. Ross, C. P., Some features of the Idaho batholith. 16th Internal. Geol. Cong. (1933) Rept., vol. 1, fig. 2, 1936. (His map has been slightly altered in places to conform with additional information appearing in various Idaho Bureau of Mines and Geology publications. The base for Plate 2 is the Idaho Bureau of Highways map for 1938. The approximate location of the gold containing areas are from Ross, C. P., The metal and coal mining districts of Idaho, with notes on the nonmetallic mineral resources of the State, Idaho Bur. of Min. and Geol., Pamphlet 57 (1941).

22. Lindgren, Waldemar, The gold and silver veins of the Silver City, DeLamar, and other mining districts in Idaho., U. S. Geol. Sur., 20th Annual Report, part 3, p. 100 (1898-99).

23. Hailey, John, Idem, p. 64.

24. Umpleby, J. B., Ore deposits in the Sawtooth Quadrangle, Blaine and Custer Counties, Idaho, U. S. Geol. Sur. Bull. 580, p. 221 (1913). (At present time the greater part of this area would be in Camas County.)

25. Ross, C. P., Geology and ore deposits of the Seafoam, Alder Creek, Little Smoky, and Willow Creek mining districts, Custer and Camas Counties, Idaho. Idaho Bur. of Min. and Geol., Pamphlet 33, p. 19 (1930).

26. Ross, C. P., Idem, p. 23.

27. Lindgren, Waldemar, 20th Annual Report, p. 190.

28. Lindgren, Waldemar, Idem, p. 207.

29. Geologic Atlas of U. S., Boise Folio, U. S. Geol. Sur. Folio 45 (1898). Geology by Waldemar Lindgren.

30. Lindgren, Waldemar, The mining districts of the Idaho Basin and The Boise Ridge, Idaho. U. S. Geol. Sur., 18th Annual Report, p. 625 (1897).

31. Ballard, S. M., Geology and gold resources of Boise Basin, Boise County, Idaho. Idaho Bur. of Min. and Geol. Bull. 9 (1924).
Lindgren, Waldemar, 18th Annual Report, p. 638.

32. Lindgren, Waldemar, Idem, Plate XC and figs. 55, 56, 57, 58, 59, 63, 64.

33. Anderson, A. L., Geology and ore deposits of the Clark Fork district, Idaho. Idaho Bur. of Min. and Geol. Bull. 12 (1930).

34. Bancroft, H. H., Idem, p. 533.

35. Kirkham, V. R. D. and Ellis, E. W., Geology and ore deposits of Boundary County, Idaho. Idaho Bur. of Min. and Geol. Bull. 10 (1926).

36. Anderson, A. L., Geology and mineral resources of eastern Cassia County, Idaho. Idaho Bur. of Min. and Geol. Bull. 14 (1931).

37. Kirkham, V. R. D., A geologic reconnaisance of Clark and Jefferson and parts of Butte, Custer, Fremont, Lemhi, and Madison Counties, Idaho. Idaho Bur. of Min. and Geol. Pamphlet 19 (1927).

38. Bancroft, H. H., Idem, p. 533.

39. Staley, W. W., Mining activity in the North Fork of the Clearwater River area. Idaho Bur. of Min. and Geol. Pamphlet 54 (1940).

40. Umpleby, J. B., Some ore deposits in northwestern Custer County, Idaho. U. S. Geol. Sur. Bull. 539 (1913).

41. Umpleby, J. B., Idem, p. 47.

42. Idem, p. 49.

43. Anderson, A. L., Geology of the gold-bearing lodes of the Rocky Bar district, Elmore County, Idaho. Idaho Bur. of Min. and Geol. Pamphlet 65 (1943).

44. Anderson, A. L., Idem, pp. 17, 20.

45. Lindgren, Waldemar, A geologic reconnaisance across the Bitterroot Range and Clearwater Mountains in Montana and Idaho. U. S. Geol. Sur. Prof. Paper 27 (1904).
20th Annual Report, p. 232.

46. Shenon, P. J. and Ross, C. P., Geology and ore deposits near Edwardsburg and Thunder Mountain, Idaho. Idaho Bur. of Min. and Geol. Pamphlet 44 (1936).

47. Lindgren, Waldemar, Prof. Paper 27, p. 83.

48. Idem, p. 85.

49. Lindgren, Waldemar, Prof. Paper 27, p. 84. (At the time of the report Pierce was in Shoshone County.)

50. Lindgren, Waldemar, Idem, p. 84.

51. Lindgren, Waldemar, 20th Annual Report, p. 233 and p. 238.

52. Anderson, A. L., Geology and metalliferous deposits of Kootenai County, Idaho. Idaho Bur. of Min. and Geol. Pamphlet 53 (1940).

53. Umpleby, J. B., Geology and ore deposits of Lemhi County, Idaho. U. S. Geol. Sur. Bull. 528 (1913).

54. Lindgren, Waldemar, Prof. Paper 27.

55. Anderson, A. L., Copper mineralization near Salmon, Lemhi County, Idaho. Idaho Bur. of Min. and Geol. Pamphlet 60 (1943).
Anderson, A. L., A prelimary report on the cobalt deposits in the Blackbird district, Lemhi County, Idaho. Idaho Bur. of Min. and Geol. Pamphlet 61 (1943).

Anderson, A. L., The antimony and fluospar deposits near Meyers Cove, Lemhi County, Idaho. Idaho Bur. of Min. and Geol. Pamphlet 62 (1943).

56. Umpleby, J. B., Bull. 528, p. 49.

57. Idem, p. 63.

58. Lindgren, Waldemar, 20th Annual Report, part 3, p. 111.

59. Bancroft, H. H., Idem, p. 534.

60. Piper, A. M. and Laney, F. B., Geology and metalliferous resources of the region about Silver City, Idaho. Idaho Bur. of Min. and Geol. Bull. 11 (1926).

61. Piper and Laney, Idem, p. 37.

62. Idem, p. 43.

63. Idem, p. 50.

64. Idem, p. 63.

65. Idem, p. 70.

66. Umpleby, J. B., Geology and ore deposits of Shoshone County, Idaho. U. S. Geol. Sur. Bull. 732, p. 125 (1923).

67. Shenon, P. J., Geology and ore deposits near Murray, Idaho. Idaho Bur. of Min. and Geol. Pamphlet 47 (1938).

68. Collier, A. J., Ore deposits in the St. Joe River Basin, Idaho. U. S. Geol. Sur. Bull. 285 (1906).
Pardee, J. T., Geology and mineralization of the upper St. Joe River Basin, Idaho. U. S. Geol. Sur. Bull. 470, p. 39 (1910).

69. Ransome, F. L. and Calkins, F. C., The geology and ore deposits of the Coeur d'Alene district, Idaho. U. S. Geol. Sur. Prof. Paper 62, p. 90 (1908).

70. Shenon, P. J., Pamphlet 47, p. 18.

71. Idem, p. 21. (The Wakeup Jim claim.)

72. Shenon, P. J., Idem, p. 40.

73. Pardee, J. T., Bull. 470, p. 40.

74. Shenon, P. J., and Ross, C. P., Geology and ore deposits near Edwardsburg and Thunder Mountain, Idaho. Idaho Bur. of Min. and Geol. Pamphlet 44 (1936).

75. Shenon and Ross, Pamphlet 44, p. 1.

76. Currier, L. W., A preliminary report of the geology and ore deposits of the eastern part of the Yellow Pine district, Idaho. Idaho Bur. of Min. and Geol. Pamphlet 43, (1935).

77. Currier, L. M., Idem, p. 17.

78. Shenon and Ross, Pamphlet 44, p. 19.

79. Hill, J. M., Notes on the fine gold of Snake River, Idaho. U. S. Geol. Sur. Bull. 620, p. 271 (1916).

80. Bancroft, H. H., Idem, p. 533.

81. Hill, J. M., Idem, p. 274.

82. Bancroft, H. H., Idem, pp. 529-535.

83. Hill, J. M., Idem, p. 277.

84. Bell, R. N., Dredging for fine gold in Idaho. Eng. and Min. Jnl., vol. 73, p. 241, Feb. 15, 1902.

CANADA

IDAHO BU

GRANITIC AR

WASHINGTON

Kootenai

BOUDARY

BONNERS FERRY

Priest Lake

River

BONNER

SANDPOINT

Pend Oreille Lake

•Clark Fork

Clark Fork R.

KOOTENAI

COEUR D'ALENE

Lake
Coeurd'Alene

Coeur d'Alene *Riv.*

•Murray

WALLACE

ST. MARIES

St. Maries R.

St. Joe *River*

BENEWAH

SHOSHONE

LATAH

MOSCOW

CLEARWATER

North

Fork

LEWISTON

Clearwater

OROFINO

•Pierce

NEZ PERCE

River

Lochsa

LEWIS

NEZ PERCE

Middle Fk.

Selway

River

GRANGEVILLE

Elk City •
Fk.

South

I D A H O

Salmon

Florence

River

REAU OF MINES AND GEOLOGY
MOSCOW, IDAHO
MAP SHOWING

REAS, GOLD, AND GOLD-BASE METAL AREAS

COMPILED AND DRAWN BY W.W.STALEY

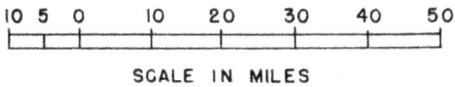

10 5 0   10   20   30   40   50

SCALE IN MILES

GRANITIC AREAS (IDAHO BATHOLITH)

INTRUDED AREAS—BELT SERIES, LAVA, BASALT, LIMESTONE, ETC.

GOLD—PLACER AND LODE

GOLD-BASE METAL—PLACER AND LODE

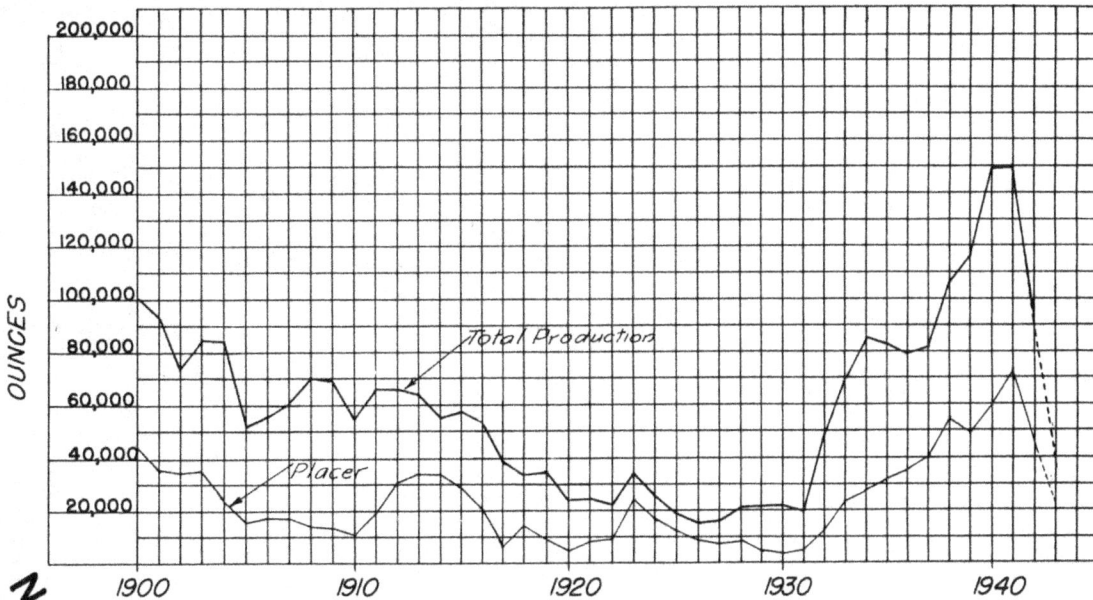

GOLD PRODUCTION—PLACER AND LODE

Total—1863—1942=8,110,004 oz. (Approx.)

Placer—1900—1942=1,026,198 oz.

Total—1900—1942=2,575,888 oz.

bbonsville

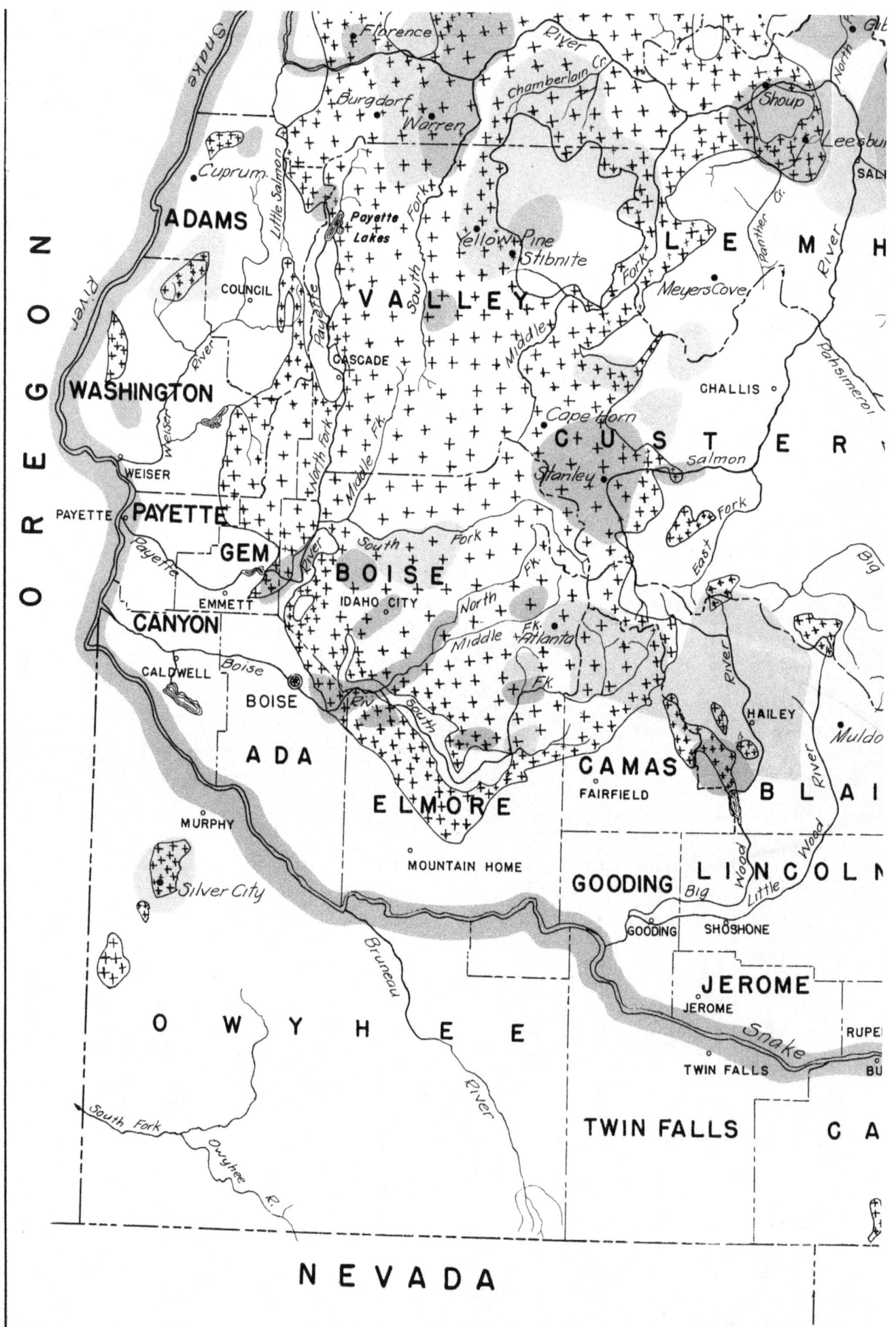

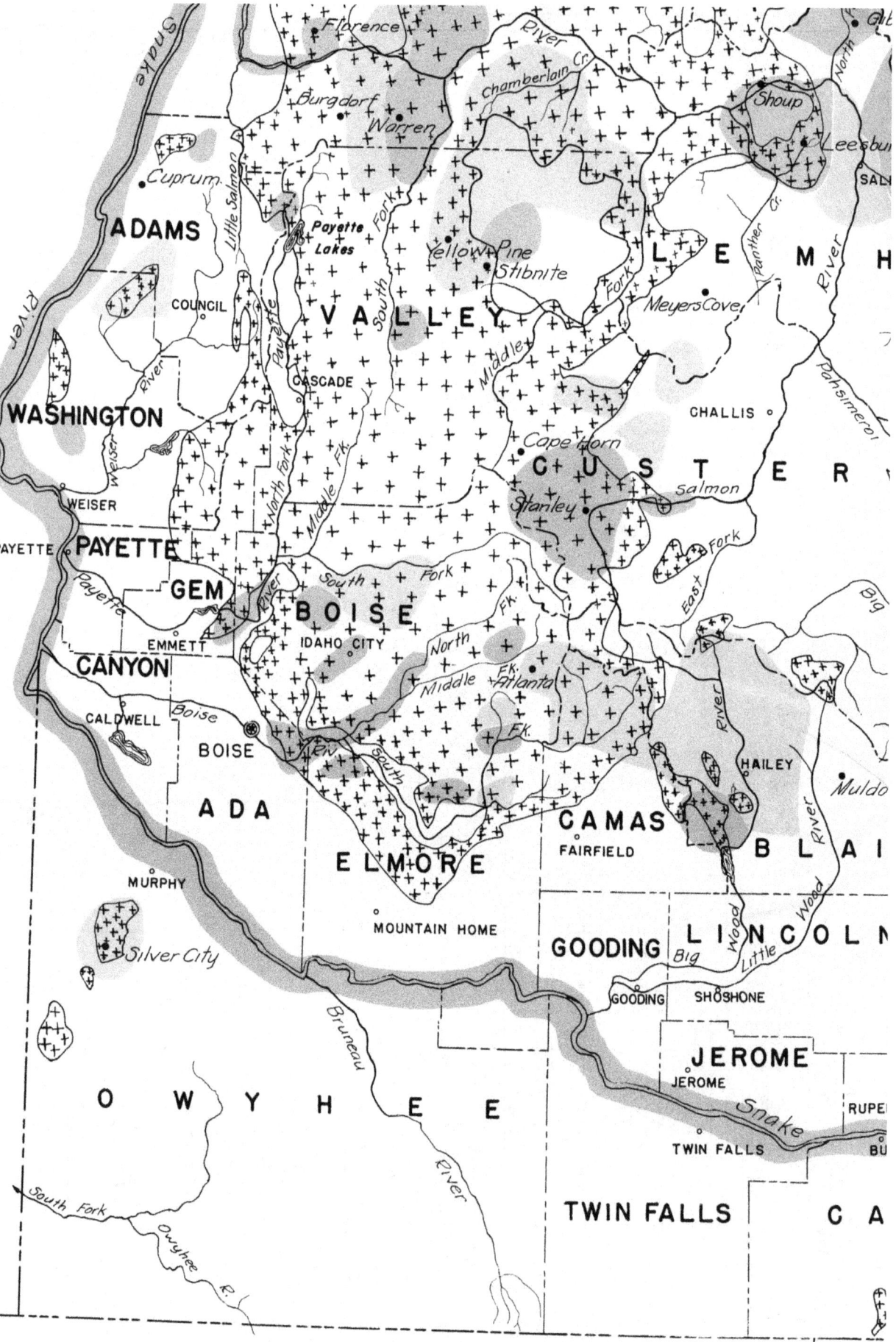

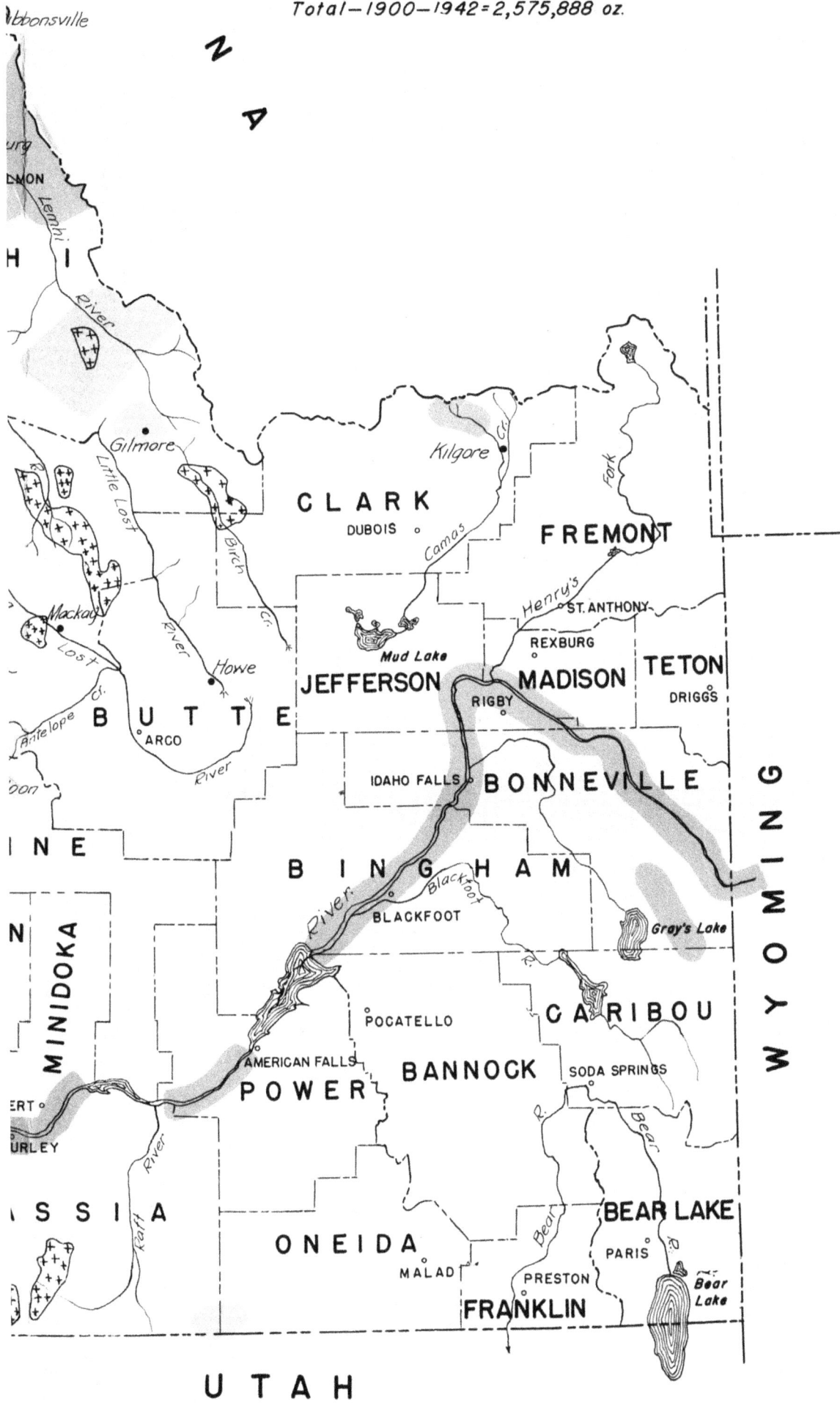

bbonsville

M O N T A N A

urg

LMON

H I

Lemhi river

Gibbonsville

Gilmore

Little Lost River

Birch Cr.

Mackay

Lost River

Howe

Antelope Cr.

on

B U T T E

ARCO

N E

Kilgore

C L A R K

DUBOIS

Camas

Mud Lake

J E F F E R S O N

Henry's Fork

St. Anthony

REXBURG

RIGBY

F R E M O N T

M A D I S O N

T E T O N

DRIGGS

IDAHO FALLS

B O N N E V I L L E

W Y O M I N G

M I N I D O K A

B I N G H A M

River

Blackfoot

BLACKFOOT

R.

Gray's Lake

POCATELLO

C A R I B O U

ERT

URLEY

AMERICAN FALLS

P O W E R

B A N N O C K

SODA SPRINGS

Raft River

SSIA

O N E I D A

MALAD

Bear R.

Bear R.

B E A R L A K E

PARIS

PRESTON

Bear Lake

F R A N K L I N

U T A H

www.ingramcontent.com/pod-product-compliance
Lightning Source LLC
Chambersburg PA
CBHW051351200326
41521CB00014B/2538